ENGINEERING TOOLS, TECHNIQUES AND TABLES

AUTOMATION SYSTEMS OF THE 21ST CENTURY

NEW TECHNOLOGIES, APPLICATIONS AND IMPACTS ON THE ENVIRONMENT & INDUSTRIAL PROCESSES

ENGINEERING TOOLS, TECHNIQUES AND TABLES

Additional books in this series can be found on Nova's website under the Series tab.

Additional e-books in this series can be found on Nova's website under the e-book tab.

MANUFACTURING TECHNOLOGY RESEARCH

Additional books in this series can be found on Nova's website under the Series tab.

Additional e-books in this series can be found on Nova's website under the e-book tab.

ENGINEERING TOOLS, TECHNIQUES AND TABLES

AUTOMATION SYSTEMS OF THE 21ST CENTURY

NEW TECHNOLOGIES, APPLICATIONS AND IMPACTS ON THE ENVIRONMENT & INDUSTRIAL PROCESSES

DOUG ARENT
AND
MONICA FREEBUSH
EDITORS

nova publishers
New York

For permission to use material from this book please contact us:
Telephone 631-231-7269; Fax 631-231-8175
Web Site: http://www.novapublishers.com

NOTICE TO THE READER

LIBRARY OF CONGRESS CATALOGING-IN-PUBLICATION DATA

Automation systems of the 21st century : new technologies, applications and impacts on the environment & industrial processes / editors, Doug Arent and Monica Freebush.
pages cm
Includes index.
ISBN 978-1-62948-262-0 (hardcover)
1. Automation. 2. Automatic control. I. Arent, Doug, editor of compilation. II. Freebush, Monica, editor of compilation.
T59.5.A837 2011
629.8--dc23

2013036359

Published by Nova Science Publishers, Inc. † New York

CONTENTS

PREFACE

In this book, the authors present current research in the study of new technologies, applications and impacts on the environment and industrial processes relating to automation systems in the 21st century. Topics discussed in this compilation include automated native sample preparation for proteome analysis; modern automation in warehousing by using shuttle based technology; flow methodologies from the point of view of green analytical chemistry; cross-coupling digital control of multi-independent servo-drives; and challenges on the way to automatic network intrusion detection.

A multidimensional chromatographic method is presented in Chapter 1 which is adapted for fractionation of body fluids for proteomic analysis and biomarker search.

This method combines native size exclusion (SEC, first dimension, 1D), followed by anion exchange (AEC, 2D) and lectin affinity (LAC, 3D) chromatography. After serial 1D-fractionation, all further procedures are performed in microplate format. Thus, beginning with 2D-fractionation, parallelization and automation is achieved for separation, spectrophotometric readout, temporary storage, hit picking, medium exchange, digest, desalting, and finally storage within an autosampler of LC-MS.

The central unit is an automated workstation enabling automated multichannel pipetting and robotic handling of microplates, reservoirs, and column arrays. The whole fractionation process, hit picking and analytics are controlled by adapted software packages. Some tools and tricks have been developed in order to improve sample preparation between fractionation steps, such as for medium exchange and desalting, re-concentration, and readout in microplates. Per working day, four serial 1D-runs or four synchronic 2D-runs can be performed including spectrophotometric readout. Thus, up to four samples may be processed comparably (1D) or in parallel (2D) within two days. Thereafter the majority of fractions are run through tryptic digest and mass spec; the few highly complex fractions remaining must be further separated by LAC before digest.

Using commercially available columns (1D), the protein input of samples is scalable from 2 to more than 100 mg and from 2 to 1000 kDa molecular weights. Globally, protein recoveries based on UV measurements from the corresponding load are 93.7 ± 2.7, 97.1 ± 11.2, and 88.8 ± 3.6% and best precisions are 3.3, 3.8, and 5.1% CV with 1D-, 2D-, and 3D-fractionation, respectively.

For process control and all-round analytics, a new software tool was developed accompanying the whole fractionation process. Its automatic interfaces collect entire data sets available including sample specification, location, and readout data of all fractions and identified proteins into a central data warehouse. Thus, immediate data access, comparison, and visualization in versatile charts and reports are possible.

Besides automation, this method has several advantages: high proteome coverage, flexible dynamic range with respect to molecular weight and sample amount, optional enzymatic and immunological analytics additional to mass spectrometry. Without exception, all intrinsic components and information are preserved after fractionation, including natural complex formation, fragmentation, and biological activities.

Such information is beneficial for comprehensive profiling, for biomarker search, as well as for efficient evaluation.

Preliminary applications show versatility with profiling plasma proteomes of humans, cattle, goat and mouse, and human cerebrospinal fluid (CSF). *E.g.*, with two human plasma samples, 3166 and 816 non-redundant proteins could be identified supported by ≥ 1 and ≥ 2 peptides, respectively.

Applying this method to biomarker search yields candidate markers that have already been validated in severe inflammation and Alport syndrome. Further application in neurodegenerative diseases and Psoriatic arthritis identifies huge numbers of altered proteins in CSF and plasma, respectively, which may present promising candidates for evaluation.

In Chapter 2, a conceptual model for calculating the system performance of shuttle based automated warehouse system (SBAWS) is presented. According to the global trends in material handling and warehousing, a SBAWS should encompass not only the installation of equipment with fastest drives, but also consider the energy consumption and environment effects on the installed equipment. The proposed conceptual model illustrates energy regeneration and throughput performance calculations of a SBAWS. Thus, the proposed model enables design of a high-throughput-performance SBAWS with reduced energy consumption through energy regeneration and so reduced CO2 emission. It is believed that by this study, through considering the throughput performance and the energy efficiency (regeneration) simultaneously a SBAWS will provide a significant contribution to the environment and the economy.

Advances in environment, agriculture or medicine raised an increasing demand for control analyses and posed various challenges to analytical chemists such as the need to develop new methods exhibiting as much sensitivity, selectivity, easy automation, low cost of equipment, simplicity, and environmental friendliness as possible. The inception of flow analysis has proved to be an excellent tool for handling solutions and, consequently, performing methods related to wet chemical analysis. In this sense, the main flow methodologies employed in recent years, in addition to conventional flow injection analysis, have been sequential injection analysis, lab-on-valve, and multicommutated flow techniques. Included in the last one, multicommutated flow analysis, multysiringe flow analysis, and multipumping flow systems are the most common. All these computer-controlled techniques have introduced a great number of advantages, mainly a dramatic decrease of the needed sample and reagents volumes, higher repeatability and miniaturization.

Chapter 3 covers a brief description about the fundamentals of the above mentioned methodologies that are normally coupled to different analytical systems, including chromatographic, spectroscopic, and electrochemical ones. A critical discussion of the advantages, handicaps and current trends of each methodology will be done, finishing with some applications in the environmental, agri-food and clinical fields.

The motion coordination in a system with multiple number of servomechanisms or axis has drawn much attention, especially in the modern manufacturing development. In early 1980s, the concept of cross-coupling control with a symmetrical structure was primarily introduced for the machine tool control. Quite original structures of cross-coupling speed- and position-control of the shafts of two distant electromechanical drives based on the application of the concept of the so-called electrical shaft were introduced and discussed by the author of Chapter 4 in mid 1990s. More concretely, the proposed cross-coupling control is accomplished by the digital simulation of stiffness and friction of a virtual twisting mechanical connection between the output shafts of the drives. Since then, a number of contributions in terms of the improvement of the suggested cross-coupling concept can be found in the literature.

In this chapter, the synthesis of the structure of coordinated control in a multi drives positioning system is presented. The proposed structure is implemented in the case of three cross-coupling servomechanisms. Also, the control strategy is experimentally verified using the system for fast development and implementation of control algorithms dSPACE R&D DS1104 that provides a comfortable work in the MATLAB®/Simulink environment.

The use of Network Intrusion Detection Systems (NIDS) emerges as a tool aimed at helping system administrators to monitor and identify computer attacks. A NIDS monitors network segments and performs analysis at different network protocol layers to identify suspicious activities.

In Chapter 5 the authors survey the most relevant works in the field of automatic network intrusion detection in the last 15 years. Their goal is to analyze the intrusion detection problem from a wider perspective, which can help to identify the causes behind the lack of acceptance of novel approaches by system administrators as well as provide new lines of work to the intrusion detection research community. In contrast to previous surveys, the authors consider other issues than just the embedded classification problem. Features such as traffic model acquisition scheme, usage frequency and adaptability are analyzed in two different stages of intrusion detection. In particular, the authors considered the initial traffic model generation and the further traffic model adjustment stages. Their research has shown that the level of human interaction is still high during both intrusion detection stages. This situation responds to the fact that some of the assumptions in which these approaches rely on, do not always hold. The availability of network traffic labeled as intrusive or normal, or the presence of attack-free network traffic, are two usual assumptions followed by many alternative techniques. Unfortunately, ensuring such assumptions demands a lot of work from security experts which is precisely what is wanted to avoid.

In: Automation Systems of the 21st Century
Editors: D. Arent and M. Freebush
ISBN: 978-1-62948-262-0

Chapter 1

AUTOMATED NATIVE SAMPLE PREPARATION FOR PROTEOME ANALYSIS

Sindy Wendler[1], Bärbel Tautkus[1], Sabine Nemitz[1], Jelena Pesek[1], Thomas Krüger[1], Stefan Opitz[1], Matthias Bartz[1], Steffen Richter[2], Heiko Oehme[2], Thomas Haenel[2], Thomas Moore[2], Stefan Kreusch[3], Benedikt Hanf[4], Lutz Schmidt[4] and Heidrun Rhode[1,*]

[1]Institute of Biochemistry I, Jena University Hospital, Friedrich Schiller University Jena, Jena, Germany
[2]CyBio-AG, Jena, Germany
[3]Scienova GmbH, Jena, Germany
[4]X-CASE GmbH, Ilmenau, Germany

Dedicated to Prof. Dr. Med. Habil. Anton Horn on the occasion of his 75th birthday. He initiated and promoted the first steps of the work presented here.

* E-mail address: Heidrun.Rhode@med.uni-jena.de; Address: Heidrun Rhode, Institute of Biochemistry I, Jena University Hospital, Friedrich Schiller University Jena, 07743 Jena, Nonnenplan 2-4, Germany. (Corresponding author)

Abstract

A multidimensional chromatographic method is presented which is adapted for fractionation of body fluids for proteomic analysis and biomarker search.

This method combines native size exclusion (SEC, first dimension, 1D), followed by anion exchange (AEC, 2D) and lectin affinity (LAC, 3D) chromatography. After serial 1D-fractionation, all further procedures are performed in microplate format. Thus, beginning with 2D-fractionation, parallelization and automation is achieved for separation, spectrophotometric readout, temporary storage, hit picking, medium exchange, digest, desalting, and finally storage within an autosampler of LC-MS.

The central unit is an automated workstation enabling automated multichannel pipetting and robotic handling of microplates, reservoirs, and column arrays. The whole fractionation process, hit picking and analytics are controlled by adapted software packages. Some tools and tricks have been developed in order to improve sample preparation between fractionation steps, such as for medium exchange and desalting, re-concentration, and readout in microplates. Per working day, four serial 1D-runs or four synchronic 2D-runs can be performed including spectrophotometric readout. Thus, up to four samples may be processed comparably (1D) or in parallel (2D) within two days. Thereafter the majority of fractions are run through tryptic digest and mass spec; the few highly complex fractions remaining must be further separated by LAC before digest.

Using commercially available columns (1D), the protein input of samples is scalable from 2 to more than 100 mg and from 2 to 1000 kDa molecular weights. Globally, protein recoveries based on UV measurements from the corresponding load are 93.7 ± 2.7, 97.1 ± 11.2, and 88.8 ± 3.6% and best precisions are 3.3, 3.8, and 5.1% CV with 1D-, 2D-, and 3D-fractionation, respectively.

For process control and all-round analytics, a new software tool was developed accompanying the whole fractionation process. Its automatic interfaces collect entire data sets available including sample specification, location, and readout data of all fractions and identified proteins into a central data warehouse. Thus, immediate data access, comparison, and visualization in versatile charts and reports are possible.

Besides automation, this method has several advantages: high proteome coverage, flexible dynamic range with respect to molecular weight and sample amount, optional enzymatic and immunological analytics additional to mass spectrometry. Without exception, all intrinsic components and information are preserved after fractionation, including natural complex formation, fragmentation, and biological activities.

Such information is beneficial for comprehensive profiling, for biomarker search, as well as for efficient evaluation.

Preliminary applications show versatility with profiling plasma proteomes of humans, cattle, goat and mouse, and human cerebrospinal fluid (CSF). *E.g.*, with two human plasma samples, 3166 and 816 non-redundant proteins could be identified supported by ≥1 and ≥2 peptides, respectively.

Applying this method to biomarker search yields candidate markers that have already been validated in severe inflammation and Alport syndrome. Further application in neurodegenerative diseases and Psoriatic arthritis identifies huge numbers of altered proteins in CSF and plasma, respectively, which may present promising candidates for evaluation.

Abbreviations

AEC	anion exchange chromatography
ALS	Amyotrophic lateral sclerosis
AUC	area under the curve(s)
ConA	concanavalin A
CSF	cerebrospinal fluid
CV	coefficient of variation
DEAE	diethylaminoethyl-
EDTA	ethylenediaminetetraacetic acid
GlcNAc	N-acetyl-D-glucosamine
IEF	isoelectric focusing
LAC	lectin affinity chromatography
α-MMP	methyl α-D-mannopyranoside
SBS	Society for Biomolecular Sciences
SEC	size exclusion chromatography
WGA	wheat germ agglutinin

1. Introduction

Proteomic techniques including biomarker discovery require considerable sample fractionation before mass spectrometric readout. These procedures are of notable complexity, enormous procedure numbers and manual labor.

To date the biomarker output passing through strict quantitative evaluation is unexpectedly, but noticeably low. In our opinion, this may result from two main possibilities: a) insufficient comprehensiveness and b) insufficient overall reproducibility of the used method combinations. This yields unreliable candidate markers. Unfortunately, both comprehensiveness and reproducibility and therefore resolution seem to exclude each other with some of the current methodologies. Thus, proteomic procedures need in particular analytical quality validation, as recently estimated [1].

One approach that could minimize manual labor and overall imprecision is parallelization of all procedures with maximal automation. Such endeavors begin with intelligent selection of fractionation and preparation principles that could be parallelized and automated, and that also cover the whole proteomes considered.

Here we present the current state of our multidimensional chromatographic method, which is adapted to fractionation of body fluids for proteomic analysis and biomarker search.

Over years this method has being developed, iteratively optimized, and already applied to biomarker search [2-7]. This method combines native size exclusion (SEC, first dimension, 1D), followed by anion exchange (AEC, 2D) and lectin affinity (LAC, 3D) chromatography. After serial 1D-fractionation, further procedures are performed in compatibility with microplate format. Thus, beginning with 2D-fractionation, consequent parallelization and automation is now achieved.

Besides automation, this method has several advantages: high proteome coverage, flexible dynamic range with respect to molecular weight and sample amount, optional enzymatic and immunological analytics additional to mass spectrometry. Without exception, all intrinsic components and information are preserved after fractionation, including natural complex formation, fragmentation, and biological activities.

2. FractionationOverviewand Methods

2.1. Overview

Principally, this method combines three native separation principles, 1D-SEC, followed by 2D-AEC and 3D-LAC. After fractionation, the whole sample content resides in several hundred liquid fractions of characteristic mixtures in microplate wells (Scheme 1).

Advantages

This combination of chromatographic methods is scalable with respect to molecular weight of constituents and starting protein amount and thus, adaptable to various sample types. The first step determines capacity. Here, commercially available columns, chromatographic systems and fractionation devices may be used that sort fractions into microplate format. Since all constituents remain in solution, liquid handling techniques may be applied throughout and therefore all techniques that have been developed to SBS standardized format. Thus, the microplate format is a precondition for standardization, parallelization and automation.

Additional to mass spectrometry, optional spectrophotometric, enzymatic and immunological analytics may be applied. Moreover, non-protein constituents are retained and may be analyzed too [2]. Along the way, an identified protein or biomarker is characterized by its location within the three-dimensional distribution space and refers attributes of native molecular weight, complex formation, fragmentation, modification, charge, and glycosylation. Since total protein yield may be controlled and any depletion is avoided the proteomes are practically covered entirely.

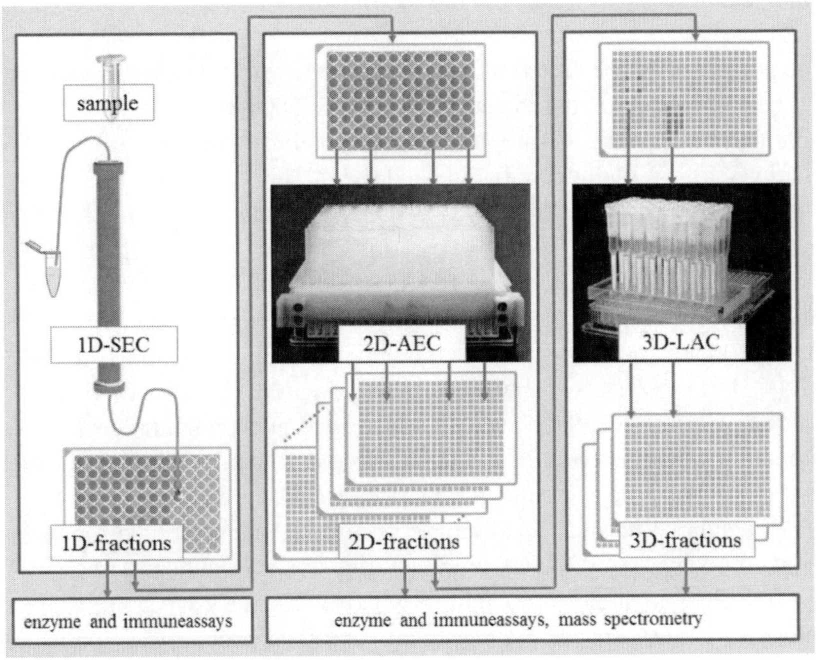

Scheme 1.

Drawbacks

With protein separation each chromatographic step results in dilution. Depending on sensitivity and volume requirements of analytics, re-concentration may be attained with *e.g.* MD100 (see section 3.2., this chapter). The automated workstation is prototypically assembled and installed. Columns and column arrays are hand- and self-made tools and are still not available commercially.

2.2. 1D-Fractionation by Size Exclusion Chromatography

1D-fractionation and analysis of the fractions from blood-derived samples were performed as described previously [2, 4]. Briefly, samples were separated on Superdex™ 200 columns (GE Healthcare, Munich, Germany) controlled by an Äkta purifier™ system (GE Healthcare). After the void volume, 48 to 96 fractions were collected into 96-well microplates or chilled 96-well deep well plates. Column calibration was performed using thyroglobulin (669 kDa), ferritin (440kDa), catalase (232 kDa), BSA (67 kDa), Chymotrypsin (25 kDa) and RNase (13.7 kDa).

The percentage of peptide fractions was calculated by Unicorn™ software (GE Healthcare) using areas under the curves (AUC) at 280 nm.

Column size depends on sample type and protein amount.

Maxi version: With human, bovine and goat serum or plasma, and pooled and concentrated cerebrospinal fluid (CSF), 1 mL of samples was injected (sample content: 0.5-0.9 mL, 20-120 mg total protein) and separated on a HiLoadSuperdex™ 200 column (16/60, GE Healthcare). To fractionate serum and pooled and concentrated CSF, 10 mMTris-HCl (Merck KGaA, Darmstadt, Germany, 1.08382.2500) was used as an elution buffer containing 150 mM NaCl (pH 7.4). For EDTA plasma separation, 1 mM EDTA (Sigma-Aldrich, Taufkirchen, Germany, E-9884) was added to prevent coagulation on the column. The running temperature was 20°C. After the void volume of 37 mL, 96 fractions of 0.5 - 1 mL were collected using a flow rate of 1 mL/min.

Midi version: With individual mouse serum, 100 μL of pooled samples were separated on a Superdex™ 200 GL column (10/300, GE Healthcare) in 10 mM Tris-HCl including 150 mM NaCl (pH 7.4) at 20 °C. Here a sample content of 50-100 μL corresponding to 3-14 mg total protein was applied. After the void volume of 6 mL, 96 fractions of 250 μL were collected using a flow rate of 350μL/min.

Mini version: With mouse serum or concentrated CSF (Vivaspin 6, GE Healthcare, MWCO 5 kDa, 28-9322-94) 12.5 μL were separated on a Superdex™ 200 GL column (5/150, GE Healthcare) in 10 mM Tris-HCl including 150 mM NaCl (pH 7.4) at 20 °C. Here samples of 100μL corresponding to 0.2-1.5 mg total protein were applied. After the void volume of 1 mL, 48 fractions of 50 μL were collected into microplates using a flow rate of 50 μL/min.

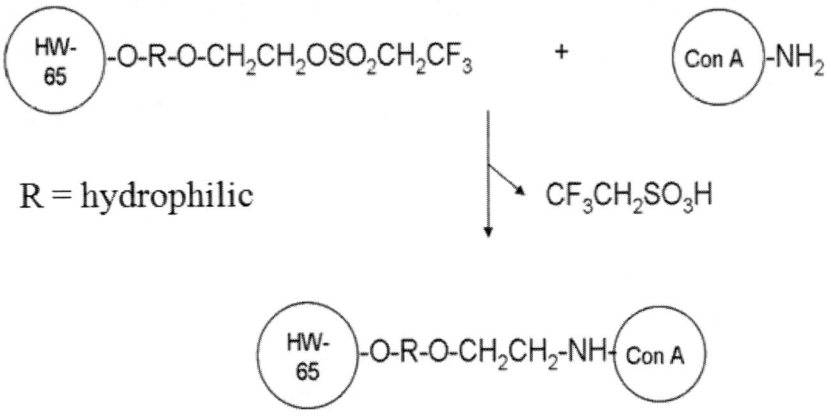

Figure 1. Immobilization principle of Con A on Toyopearls AF-Tresyl-650M.

2.4.2. 3D-Separation Conditions

Generally, prior to use, lectin columns (see 2.4.1. and 3.1.2, this chapter) were equilibrated with binding buffer (20 mM Tris/HCl pH 7.4 containing 150 mM NaCl, 1 mM $CaCl_2$, and 1 mM $MnCl_2$) and adjusted to one partition of a 384-well UV-Star microplate applying an adapter (see Figure 3). After sample application the columns were washed with binding buffer (I). Thereafter, bound proteins were successively eluted by 0.5 M α-MMP (II), and 0.2 M GlcNAc (III) in binding buffer, and by regeneration buffer (IV, 20 mM Tris, 500 mM NaCl, pH 8.0).

Thus, four different 3D-fractions were produced: I, proteins without affinity to the lectins; II, proteins containing mannose-rich glycans; III, proteins with gylcans containing glucose and GlcNAc; and IV, proteins with other modifications of high affinity to the ligands. The maximum fractional volume is determined by the capacity of the wells of the microplates used, ~100μL.Since complete fractional elution needs at least 400 μL, four aliquots (100 μL) of each 3D-faction were produced in microplates. Alternatively, the eluates may be pooled in vessels with higher volume capacity.

2.5. Readout

With chromatographic fractions in microplates versatile spectrophotometry analysis may be achieved applying the facilities of reader technologies. With our studies we used measurements of absorbance at 205, 215, and 280 nm in order to quantify protein concentration on different sensitivity levels [10], at 340 nm to quantify some metabolites such as nucleotides, at 460 nm for determination of chromophores such as heme and bilirubin, and at 700 nm for the estimation of blanks and turbidity of each microplate well. Absorbance of aliquots of 1D-fractions (150 μL) and entire 2D- and 3D-fractions were read in UV-Star microplates using a SpectraMax Plus[384] (Molecular Devices, Sunnyvale, CA, US). Moreover, protein composition may be fluorimetrically characterized, *e.g.*, using selective fluorescense signals of amino acids, modifications and advanced glycationendproducts (AGEs) [8]. Fluorescence was read in black microplates (Greiner Bio-One, No. 655076) using a SpectraMax Gemini (Molecular Devices).

2.6. Digestion

Following this, 75 μL of selected 2D- or 3D-fractions were digested in 96-well twin.tec PCR plates (Eppendorf AG, Hamburg, Germany, 0030128.648) according to [2].

Pooled 3D-fractions were dialyzed against deionized water (MD100, see section 3.2, this chapter), concentrated (Speedvac, Bachofer GmbH, Reutlingen, Germany), and dialyzed again against 20mM NH_4HCO_3 before digestion in order to diminish sugar concentration and volume. Denaturation was performed with guanidine-HCl (Fluka, Sigma-Aldrich, 50940), final concentration 2 M) for 20 min at 90°C using a thermocycler (TB1 Thermoblock, Biometra GmbH, Göttingen, Germany). After cooling down on ice, proteins were reduced by dithiothreitol (SERVA Electrophoresis GmbH, Heidelberg, Germany, 20710, final concentration 10 mM) at 37°C for 1 h and S-carboxyaminomethylated with iodoacetamide (SERVA, 26710, final concentration 50mM) in the dark at room temperature for 30 min. Salts and reagents were removed by dialysis against 2 L 20 mM NH_4HCO_3 using MD100 (Scienova GmbH, Jena, Germany) for 1 h (see section 3.2., this chapter, [11]). Thereafter, proteins were digested by freshly dissolved bovine trypsin (about 1/40 of the protein content, Sigma-Aldrich, T1426). Samples were incubated at room temperature for 16-18 h. The reactions were stopped by formic acid (Merck, 1.00264, final concentration 10 mM) and samples were stored in microplates at -80°C until analysis by mass spectrometry.

2.7. Mass Spectrometry

Microplates containing digested fractions were re-sealed with a Platesealer slide (Greiner Bio-One, 676070), slit once per well by a scissor blade, and applied to an Accela™Autosampler (Thermo Fisher Scientific, Waltham, MA, US). Liquid chromatographic separation prior to mass spectrometry detection was performed on a Hypersil Gold UHPLC column (50×1.0 mm, 1.9 μm) with an Accela 1250 UHPLC system (both Thermo Fisher Scientific) using binary gradient elution of the mobile phase, *i.e.* 0.1% formic acid in (A) water and in (B) acetonitrile at a flow rate of 150μL/min, throughout (0-1 min 5% B, 21 min 30% B, 24 min 40% B, 25-26 min 90% B, 26.1-30 min 5% B). Tandem mass spectrometry (MS/MS) measurements were carried out on a LTQ Orbitrap Discovery (Thermo Fisher Scientific) by positive heated electrospray ionization (H-ESI) at a vaporizer temperature of 200°C. Sheath gas flow (30.0) and auxiliary gas flow (10.0) were used to dry the ion spray (both nitrogen gas flows in arbitrary units). The ionization voltage and the temperature of the ion transfer tube were set to 4.5 kV and to 275°C, respectively. The MS/MS system operated in data-dependent TOP10 mode using 1 microscan.

For this purpose, ions have been monitored in the LTQ ion trap in full scan centroid mode at m/z 350-1700. The ten most intensive ions run through collision-induced dissociation (CID) for further orbitrap high resolution (30,000) analysis (profile data type). Wideband activation was used. The automatic gain control target value for the orbitrap mass analyzer in full scan mode was 1.0 x 106. All measurements were performed with the lock mass option (m/z: 445.120025) for internal calibration. The LC-MS/MS operated via graphical interface of the Xcalibur software 2.1. Fractions were run in duplicates. After each duplicate a blank (*i.e.*, water sample) was applied to avoid carryover. For quality control and stability check a digest of 10 μM transferrin (human (Holo), SERVA Electrophoresis GmbH, 36756) was analyzed four times per microplate. Analysis was done with SIEVE 2.0, Proteome Discoverer 1.3 and Sequest database search algorithm applying a false discovery rate of <1%.

3. Newly Developed Preparation Tools

3.1. Micro Column Arrays

To date, none of the commercially available columns meet our requirements regarding sample volumes, binding capacity, flow kinetics, geometry suitable for 384-well microplates, and precision. Thus, columns and column arrays have been established and optimized by us with the help of the Central Scientific Workshops of Jena University Hospital.

3.1.1. 2D-Columns

Solid phase cartridges (prototyped by CyBio AG) were modified with the tips of pipet tips (Figure2, BRAND GmbHandCo KG, Wertheim, Germany, 702504). Porous PVC frits [12] were applied to border resins as previously described [4].

Six weak anion exchange matrices with similar particle size (30-75 μm) were tested for binding capacity, resolution and peak capacity, protein yield, free flow elution speed, stability, reusability, and precision: ACCELL Plus QMA (WAT010742) and Oasis WAX (both from Waters GmbH, Eschborn, Germany, 186002502), Toyopearl Giga Cap Q-650M (21854), Giga Cap DEAE-650M (trade sample), Super Q-650M (43205), and DEAE-650M (43201, all Tosoh Bioscience GmbH).

Out of these only one, Toyopearl DEAE-650M, showed characteristics suitable for native 2D-fractionation of plasma proteins.

A 33% (v/v) suspension of this resin in 2D-storage buffer (10 mM Tris/HCl, pH 7.4, containing 150 mM NaCl and 0.02 % (w/v) NaN_3) was prepared and devolatilized by ultrasound. 300 μL of the suspension were deposited on the top of cartridge prefilled with the lower frit and degassed 2D-storage buffer. Packaging was accomplished under continuous flow and buffer loading. Thereafter the upper frit was applied. The prepared columns with a ~100 μL gel bed were stored at room temperature. The binding capacity was determined using BSA: 4.02 mg/100 μL packed resin.

3.1.2. 3D-Columns

Pipet tips (250 μL, CyBio AG, OL3800-25-559-N) were used as cartridges in order to meet volume and kinetic requirements of available starting material and affinity separation, respectively (Figure 3). Porous PVC frits (see 3.1.1.) were applied to border resins. A 33% (w/w) suspension of the affinity-resin (see section 2.4.1.) in 3D-storage buffer was prepared and devolatilized by ultrasound. 210 μL of the suspension were deposited on the top of pipet tip prefilled with the lower frit and degassed storage buffer. Packaging was accomplished under continuous flow and buffer loading. Thereafter the upper frit was applied. The prepared columns with ~70 μL gel bed were stored at 4°C.

All columns are stacked to microplate format by appropriate adapters suitable for robotic or manual handling using multichannel pipettes, see Figures 2 and 3.

Figure 2. AEC columns. Array of 96 AEC columns (left) within the adapter for automatic processing and adjusted to a 384-well microplate and two individual columns (right).

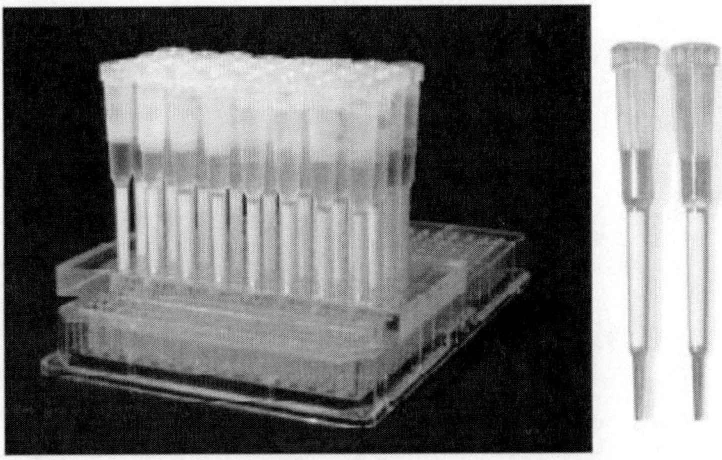

Figure 3. LAC columns. Array of 32 LAC columns (left) within the adapter for parallelized manual processing and adjusted to a 384-well microplate and two individual columns (right).

3.2. Parallel Micro Dialysis

For efficient medium exchange two variants of tools have been developed for parallel dialysis. This sample preparation principle produces compatibility to subsequent preparation steps or analysis, such as i) lowering salt concentration of fractions after 1D-SEC prior to 2D-AEC-fractionation, ii) removal of reagents used for denaturation, reduction, and modification prior to tryptic digest (see section 2.5., this chapter), and iii) desalting prior to mass spectrometry in 2D- and 3D-fractions.

Since no dialysis tool was available that meets our requirements regarding parallelization, sample volumes, volumes of outer fluid, and molecular weight cut-offs, we developed these tools ourselves.

For i, a plate comparable with a microplate is used, but with bores of 500 μL inner volume instead of wells. The lower openings were closed by a semipermeable membrane (Merck Millipore, Schwalbach, Germany, VCWP14250, [2, 13], see Figure 4). For ii and iii, U-like capillaries (MD100, Scienova GmbH) were applied as described earlier [11] and shown in Figure5. We have equipped these tools with an adapter and tightening arrays to facilitate parallel handling (Figure6).

Application note: MD100 may be also used for sample re-concentration as outlined previously, simply by evaporation [14].

Figure 4. Self-made dialysis tool with 96 bores, view to bottom with the fixed dialysis membrane.

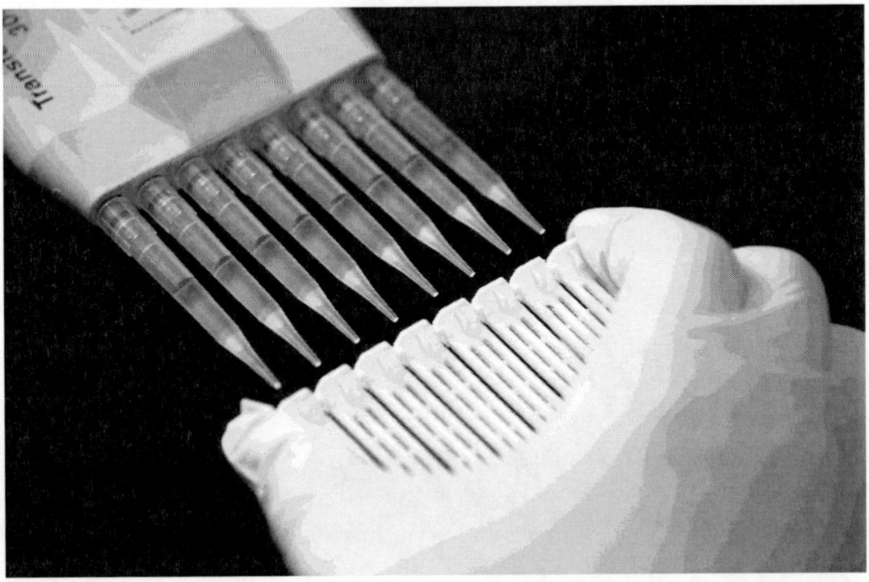

Figure 5. Parallel load of an array of eight micro dialyzers MD100.

Figure 6. Parallel retrieval of dialyzed fractions from 48 micro dialyzers MD100.

4. Automated Workstation

4.1. System Overview

Sample fractionation and the selection of fractions are performed by a fully automated workstation provided by CyBio (Figure 7, www.cybio-ag.com).

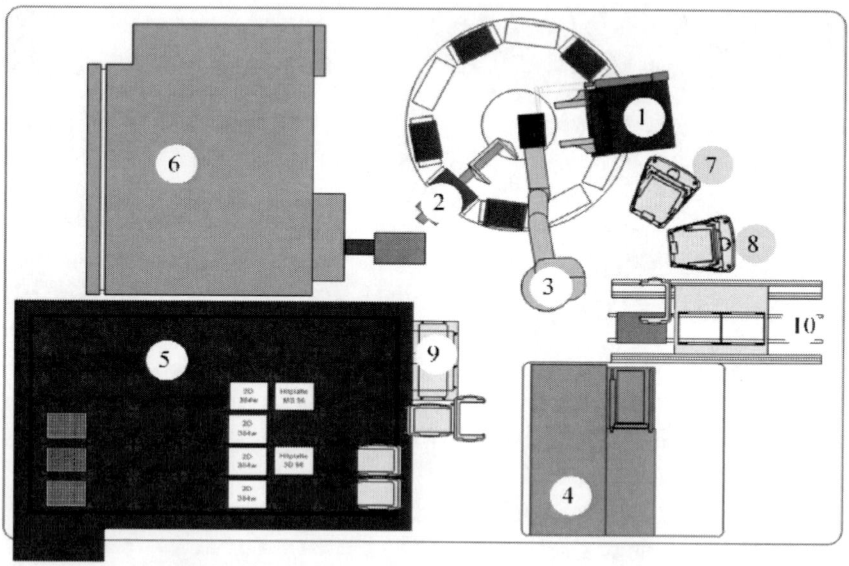

1: 96-well pipettorCyBi®-Disk. 2: Pneumatic plunger. 3: Four axis SCARA robot. 4: Absorbance reader. 5: Eight channel pipettorCyBi®-Robospense. 6: 4°C incubator with plate transfer unit. 7: Random access pipetting tip storage. 8: Storage for elution buffer reservoirs. 9: Temporary microplate storage. 10: 384-well microplate and lid storage.

Figure 7. Automated workstation system overview.

4.2. Sample Fractionation

The key element of this automated workstation is the pipettor CyBi®-Disk (Figure 8). This pipettor is equiped with 96 simultaneous pipetting channels and accomplishes the transfer of 1D-fractions and elution buffers into chromatography columns arraysarranged in 96-well microplate compatible format.

To fulfill the various liquid handling requirements for sample fractionation, the CyBi®-Disk offers a wealth of adjustment options. These include eligible pipetting speeds and immersion depths for aspirating and dispensing liquids, appropriate speeds for movement into / out of liquids, delay times before aspirating and after dispensing liquids, and variable XY positioning of pipetting tips within wells, as well as tip touch of pipetting tips to microplates and pre-wetting of tips.

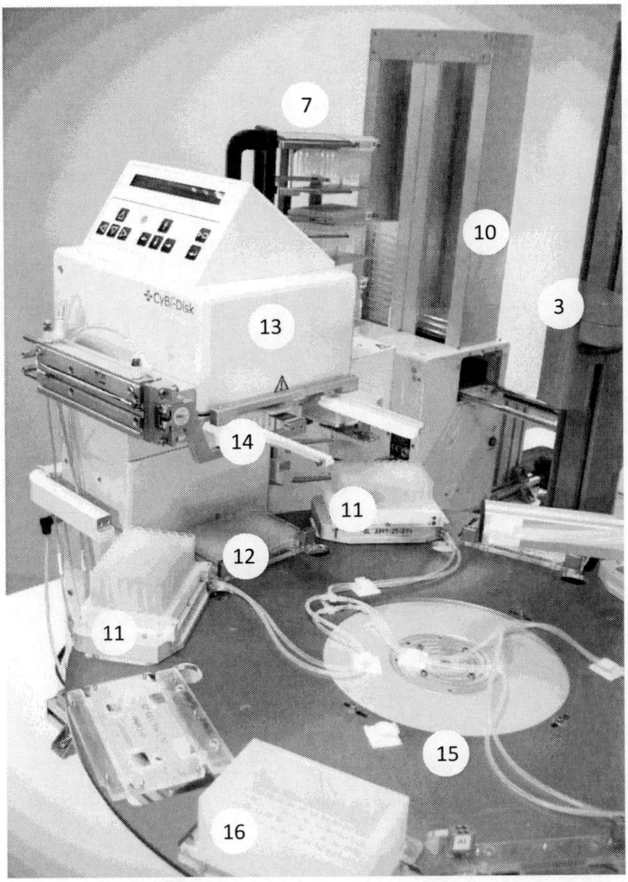

3, 7, 10: see Figure 7. 11: Column array. 12: Microplate to collect fractions. 13: 96-well pipettor head. 14: Automatic tip changer. 15: Ten place circular shelve. 16: Reservoir for buffer solution.

Figure 8. CyBi®-Disk.

The circular shelve holds up to ten microplates and/or reservoirs and provides space for various accessories. Two trays are used to process 1D-fractions from one sample. One place is required for elution into the target

(= elution) microplate. The second place is used as a storage position of the column array during target microplate replacement. Thus, up to four samples may be processed simultaneously. The remaining two places are taken by reservoirs providing buffer solutions.

Because up to four eluates (respective 2D-fractions) per 1D-fraction are collected, the column arrays must be re-adjusted to the corresponding partition of a 384-well microplate by the SCARA robot (see below) before spotting.

To ensure a reproducible flow-through of the chromatography flow, the automated workstation is equipped with a pneumatic plunger (Figure 9). After liquid transfer the pneumatic plunger applies a gentle, time limited pressure stimulus to the chromatography columns. Pressure and duration of the stimulus can be adjusted. Design and flexible mounting of the plunger ensures that each chromatography column receives an identical pressure stimulus.

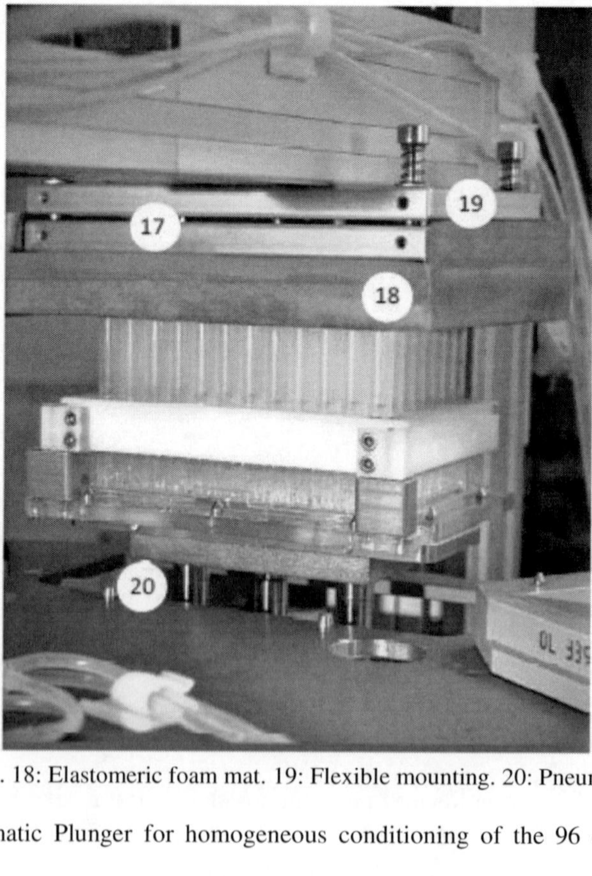

17: Pressure plate. 18: Elastomeric foam mat. 19: Flexible mounting. 20: Pneumatic cylinder.

Figure 9. Pneumatic Plunger for homogeneous conditioning of the 96 chromatography columns.

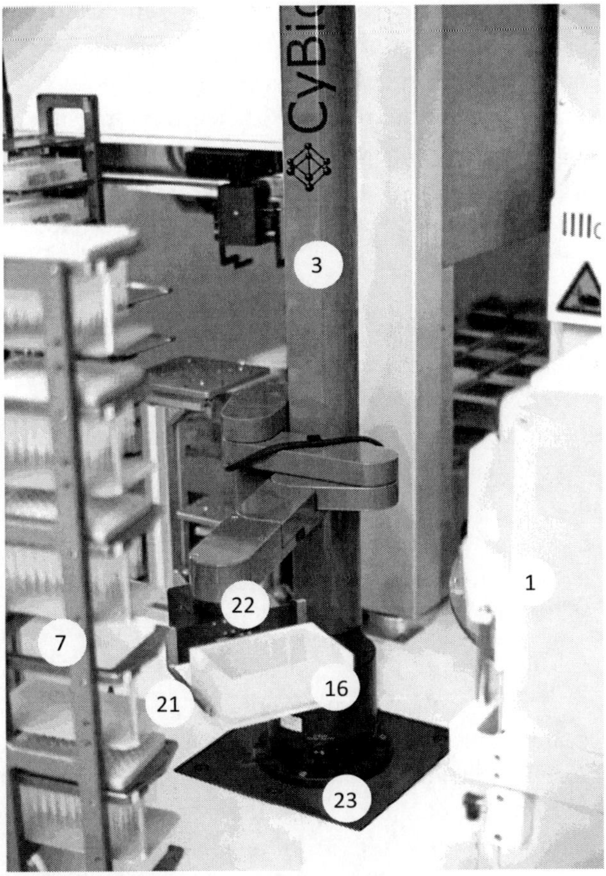

1, 3, 7, 16: see Figure 7 and 8. 21: Servo gripper. 22: Gripper collision detection. 23: Robot collision detection.

Figure 10. Four axis SCARA robot for handling various labware in the automated workstation.

A four axes SCARA robot (Selective Compliance Assembly Robot Arm) handles various labware such as deep-well plates containing 1D-fractions, target plates for 2D-fractions, tip trays and reserviors of buffer solutions (Figure 10). This robot has an operating distance of 422 millimeters (16.6 inches) and a nominal payload of 2kg (4.4 lbs.). The gripper with integrated servo technology can handle microplates in a landscape as well as portrait direction without reconfiguration.

The gripper detects missing microplates and signals an error in this situation. A possible cause might be a user oversight when setting up the system. This feature avoids pipetting from or into non-existing microplates. Moreover, the robot has an integrated collision detector, which prevents damage of labware or parts of the workstation. After elution the SCARA robot transfers the microplates containing 2D-fractions to the reader for readout, to the incubator, or to the CyBi®-Robospense (see below) for further processing. The system is able to recover from errors seamlessly once the cause has been corrected. This minimizes loss of data and samples.

Readout data are used to reduce the number of fractions for mass spectrometric analysis and to keep the overall throughput of the proteome analysis on a biologically and economically reasonable rate. Absorbance data are automatically transferred to a data warehouse (see section 5, this chapter). Based on this data, fractions are selected for the mass spectrometric analysis. The list of the selected fractions is automatically transferred back to the instrument control software for pooling.

4.3. Picking Fractions of Interest

For this, besides the pipettor CyBi® -Disk the workstation is equipped with an additional pipettor CyBi®-Robospense (Figure 11). This flexible pipettor has eight, separately controllable, equidistantly spreadable pipetting channels.Each of these 8 pipetting channels can detect the liquid level in the corresponding wells. A rich set of pipetting options comparable to those of the CyBi®-Disk (see above) ensure optimal liquid handling results. Additionally, variable indexing of wells in the X- and Y- direction is applicable for plate pooling and reformatting. Using the CyBi®-Robospense hit picking and pooling into 96-well microplates is accomplished based on the above mentioned list. All microplates containing 2D-fractions, reformatted plates or collected hits may be temporarily stored within the incubator at 4°C (39°F) before further processing for mass spectrometric analysis.

4.4. Software

The automated workstation is controlled by CyBio's own automation software, CyBio® Composer (see Figure 12). With this graphical user interface an operator can define workflows and execute them on the workstation.

24: Eight pipetting channels. 25: 35 place workspace. 26:Microplate with 2D-fractions. 27: Target plate for tryptic digest.

Figure 11. CyBi®-Robospense with deck layout for fraction picking.

The CyBio® Composer package supports sequential and parallel execution of instrument activities. Therefore it is possible to operate more than one sample at the same time and to handle labware during fractionation steps. The parallel execution ensures that the fractionation is processed as soon as possible to reduce time dependent effects.

The CyBio® Composer script editor allows quick and easy definition of workflows. The integrated adaptable error handling ensures that, even in complex error situations, the user can react in an adequate manner to avoid loss of samples. Even after breakdown of the entire system due to some error condition, the software recovers from any incomplete task continuing the workflow where it has stopped. Relevant data shared between the data warehouse of the "Biomarker Analyzer" (see section 5, this chapter) and CyBio® Composer is stored in a SQL Server database. All activities will be logged and time stamps are associated within an audit trail. Therefore all executed process steps can be retraced any time. The CyBio® Composer user management allows the restriction of access to almost every part of the software, to ensure that only authorized personal can use the appropriate features.

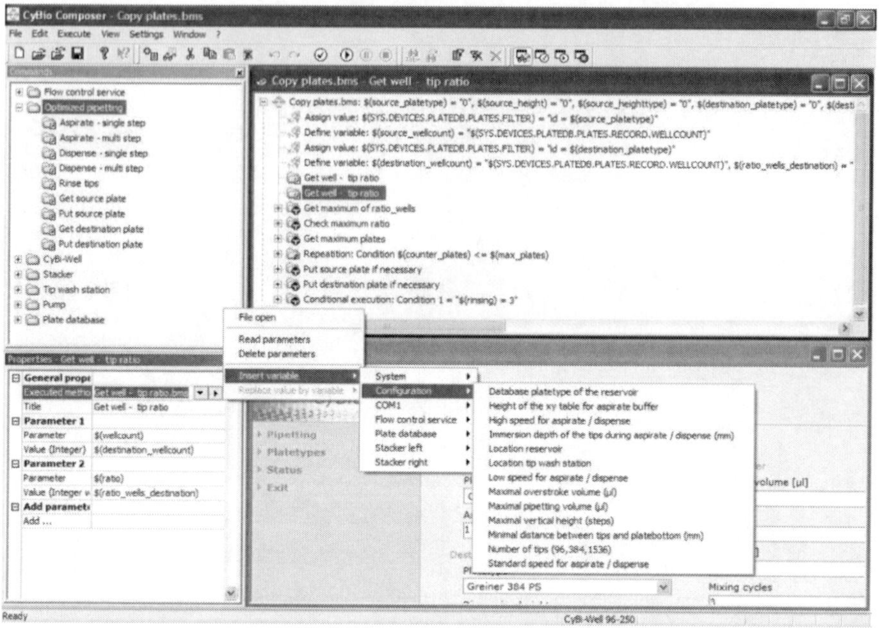

Figure 12. CyBio® Composer Software: workflow.

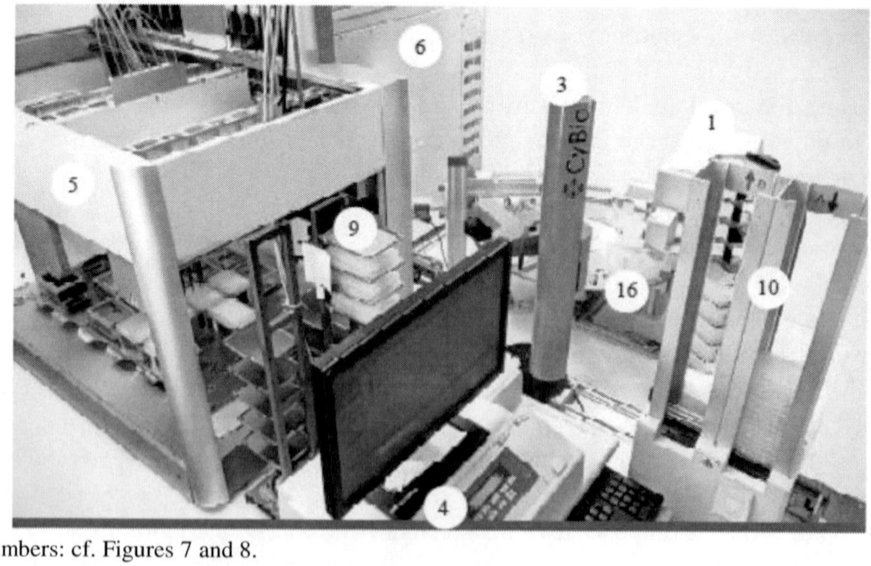

Numbers: cf. Figures 7 and 8.

Figure 13. Automated workstation overall overview.

5. Process Control and Analytics

The software package "Biomarker Analyzer" (X-CASE) handles data on samples and processes with versatile facilities for analyses, visualization and process-control.

The package consists of a computer assisted sample inventory connected to components of lab equipment by automatic interfaces, a biomedical data warehouse, and comprehensive tools for control and analysis (Figure 14). It uses business intelligence technologies like a data warehouse with reports using ad-hoc pivot-tables, graphical charts and data mining.

5.1. Sample Inventory – Lab Management for a High Throughput System

To keep track of the numerous lab objects used during high throughput workflows, all information about labware and their locations are saved automatically by the software.

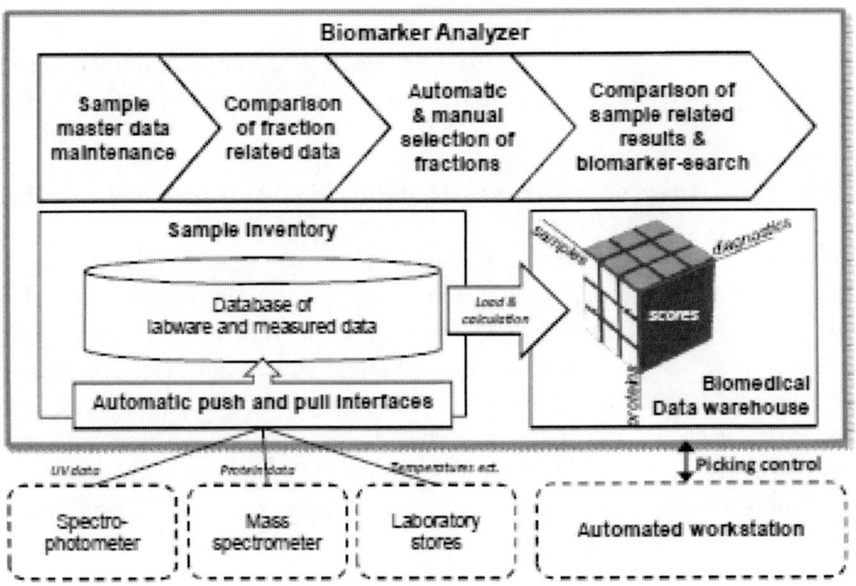

Figure 14. Components of the Biomarker Analyzer software.

This database not only locates lab objects in stores, it also links data on labware with measured and context sample data (dates, volumes, processing times, temperature, spectrophotometric readout, mass spectrometry).

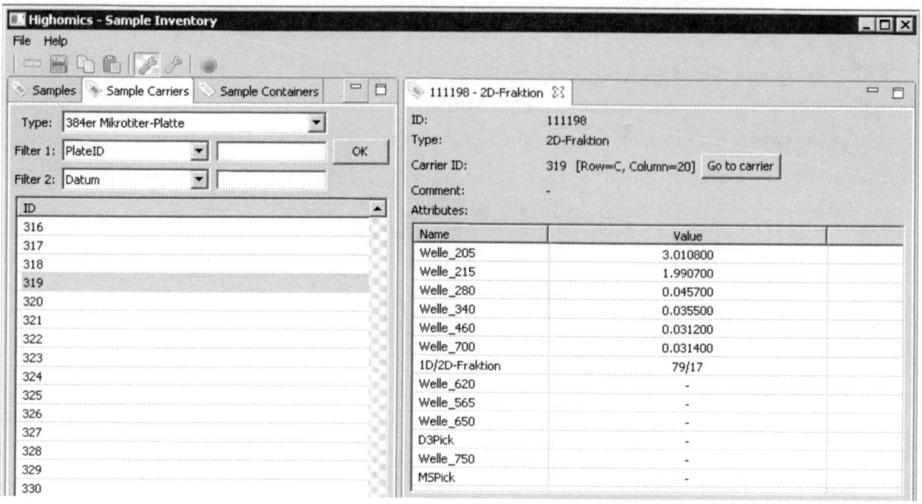

Figure 15. Sample inventory with measured data from readout.

A frontend is provided to visualize and search the laboratory logistic database (Figure 15). This lab logistics system is the base for analyses and process views.

5.2. Software Interfaces for a Fully High Throughput Automation

A robust IT-infrastructure and several automatic software interfaces were developed. Thereby, an automatic input of spectrophotometric readout and context data into the sample inventory is achieved by a pull interface. For the context data, the sample inventory determinates labware and links labware and measured sample data (see section 5.1., this chapter).

Mass spectrometric raw data are stored on a central server system with special file naming conventions. From there the pull interface generates links to the files in the sample inventory. Thus, raw data from readouts are protected from loss. Moreover, data access by links enables simultaneous processing by several scientific software tools and clients.

The spectrophotometric readout is visible instantly due to the close processing of data.

At this point, fractions may be selected already for 3D-fractionation and/or mass spectrometry. For this, selected hits are committed to the CyBi-Robospense® (see section 4, this chapter) via another interface.

Additionally, the CSV-input for the Xcalibur-software is provided to facilitate scheduling of mass spectrometry. Moreover, heavy process deviations may be identified just after readout. This all saves enormous time and vast manual labor.

5.3. Comparison and Selection of Fractions for Picking

The spectrophotometric readout of 2D-fractions is the earliest opportunity for the selection of fractions for analyses and/or further sub-fractionation. The results of the spectrometric readout, either as absorbance or calculated protein concentration are displayed using a two- or three-dimensional matrix (Figure 16) in versatile charts and reports. The software provides multiple resizable windows for visual comparison of samples and controls. The selection of fractions of interest may be performed manually by mouse click or automatically by pre-selected criteria (see section 5.2, this chapter), *e.g.*, ratios of protein concentrations or UV-absorbancies of homologous fractions, trends of neighboring fractions, and threshold values.

Moreover, alternatively, comparison may address two samples, two sample groups (disease vs. control), or multiple stages along the course of the disease.

5.4. Links/Analytical Facilities

All, data derived from spectrophotometric readout (absorbance, total protein concentration), from mass spectrometry (ID-numbers, peptide number, coverage, intensities, scores…), clinical data from patients, and data on labware and fractionation performance are linked persistently. These data are viewable by clicking on fraction identifiers (Figure 17) or by application of separate ad-hoc analysis tools. So, for deep proteomic analyses, the software provides an ad-hoc reporting system with filtering and sorting functions of all types of data, from data on individual clinical characteristics via data under fractionation to proteome data.

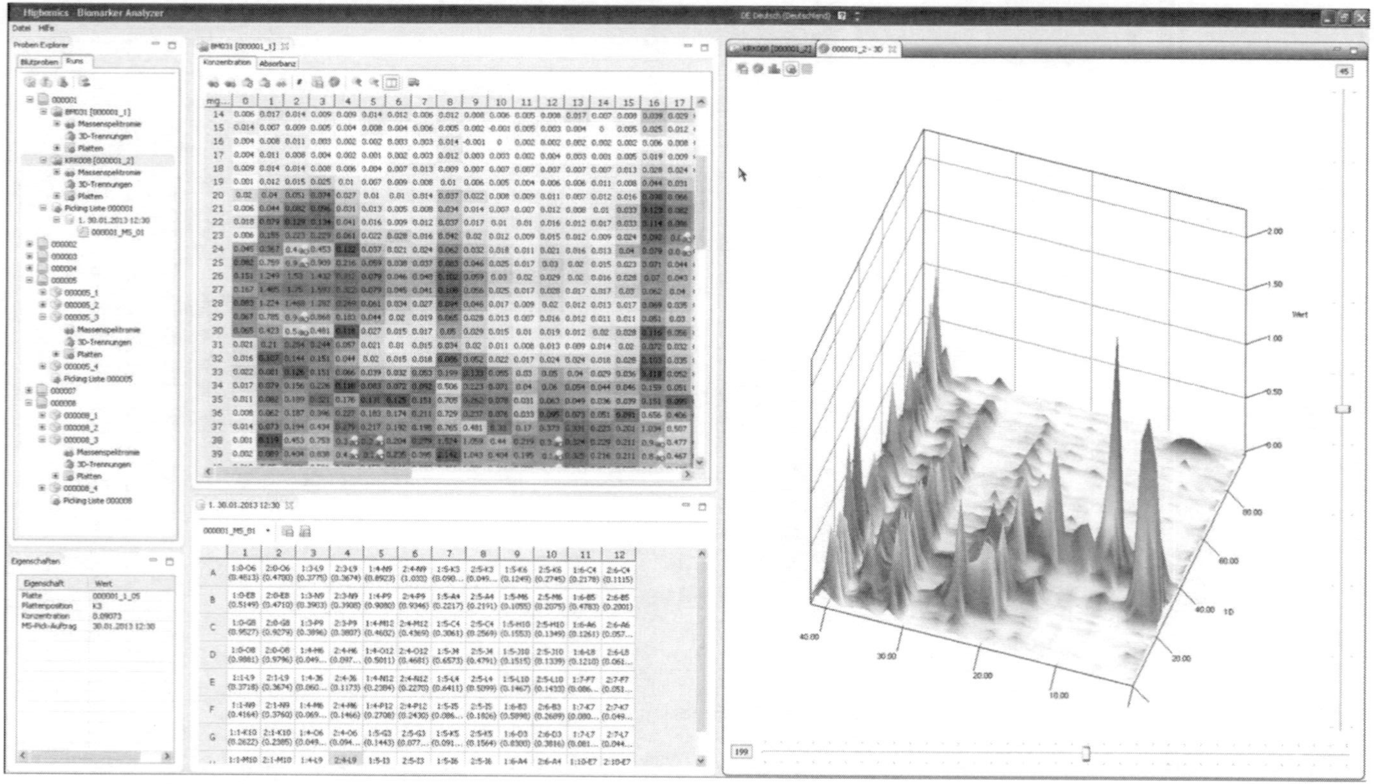

Color coded protein concentrations in 2D- (upper left) or 3D-matrix (right), and a preview of selected fractions in a picking list (bottom left).

Figure 16. Visualization of readout data.

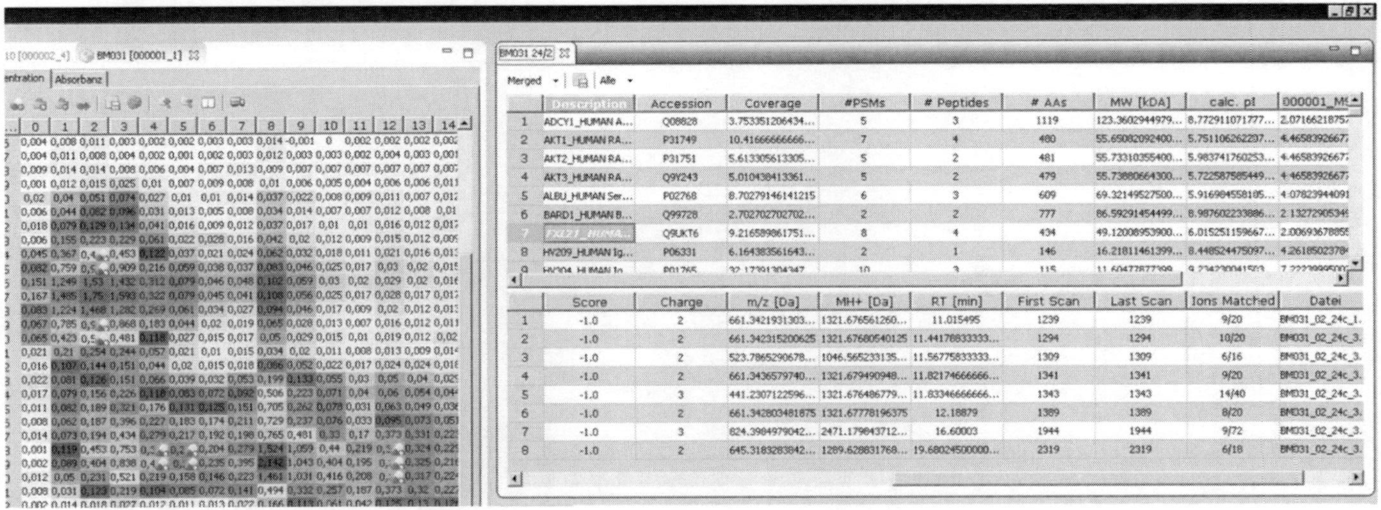

Selected fractions are marked by a pill.

Figure 17. Visual linkage of proteomic data to spectrophotometric readout of selected 2D-fractions.

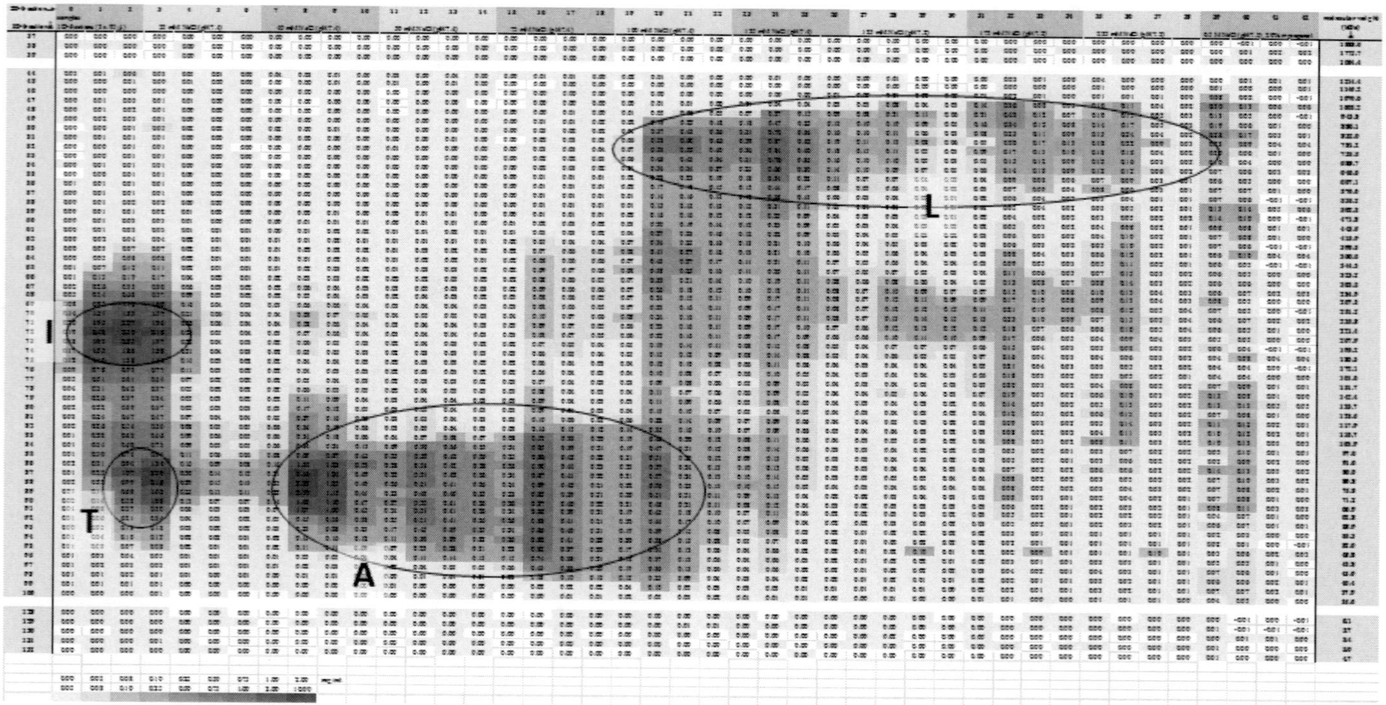

Total fractional concentrations of proteins were calculated from A_{280}–A_{700} (see section 2.5, this chapter) and are given by values (mg/mL) and grey scale; circles indicate areas rich in immunoglobulins (I), transferrin (T), albumin (A), and lipoproteins (L).

Figure 18. Distribution pattern of proteins after 2D-fractionation within 4128 fractions.

5.5. Highly Flexible and Open to Other Scientific Fields

Because research processes are changing rapidly, the software is kept flexible for integration into other systems and platforms (Windows, Linux, Macintosh) based on newest and reliable open-source-technologies, which are running on most computers. The current software was developed in Java and uses a central MySQL-database. The desktop frontend is an Eclipse Rich Client Platform (RCP), which is more interactive for analytics rather than a web-application. Additional components can be developed and installed as plugins. The software's architecture itself is separated in small main- and sub-program modules with defined interfaces. These technologies and modules are modified independently or replaced to adapt to new tasks.

To enable integration into other systems, interfaces may be developed with the integrated APIs (Application Programming Interfaces) and the frontend user may export practically all data into Excel©. Also standardized push and pull interfaces to laboratory equipment enable flexibility for future applications and opens the software to other research areas.

6. Validation Characteristics

6.1. Separation Resolution/Profiles

During native chromatography single protein entities are never completely separated from each other. Rather, proteins elute in groups according their native characteristics and interactions. Hence, with SEC, observed molecular weights reflect natural heterogeneity regarding posttranslational modifications and complex formation analogous to *in vivo* conditions. Groups of proteins or protein complexes with similar charge and similar lectin affinity co-elute with AEC and LAC, respectively. Moreover, each individual protein, complex, or fragment shows its own Gaussian distribution over a few fractions. Nevertheless, each final fraction contains its unique mixture and exhibits a much lower complexity than the original sample suitable for mass spectrometry. Generally, protein distribution after native SEC depends on column configuration and running condition. With serum and plasma, three main protein fractions were obtained under the conditions used. These peaks showed typical molecular weights of and contained lipoproteins (peak I), immunoglobulins (peak II) and albumin (peak III) as verified by mass spectrometry and immunological measurements [2, 5].

These proteins form complexes or simply co-elute with many other plasma constituents. With the maxi and midi versions (see section 2.2., this chapter) resolution was alike. In Figure 19 the typical elution profile is shown. A similar distribution but with lower resolution was achieved with the mini version.

After AEC and LAC, protein isoforms differing in their posttranslational modifications become visible. Beginning with AEC, proteins are inevitably diluted and thus native complexes may dissociate. Nevertheless, information on native molecular weight is irreversibly conserved within 1D-location of all subsequently produced sub-fractions.

A typical 2D-distribution pattern of the plasma proteome after AEC is shown in Figure 18. Highly abundant proteins, immunoglobulins, transferrin and albumin are easily visible as spots. Additionally, a simple quality control was introduced for SEC and AEC by addition and tracing phycobilins in order to check proper performance [15].

6.2. Fractionation Precision and Recovery

UV absorbance trace (1D) and high-throughput UV measurements in microplates (1D-3D) were used for quantitation of total protein concentrations and amounts [10]. This measuring principle avoided sample loss, labor and imprecision introduced by the various protein determination procedures. Moreover, the probable inclusion of UV absorbing constituents other than aromatic amino acids of proteins is beneficial with respect to biomarker search.

6.2.1. 1D, Size Exclusion Chromatography

For validation, SEC was performed 16 times within 6 days using a pool of EDTA-plasma from two healthy staff members (female, 27 and 30 years, see section 2.2., this chapter). Aliquots (1.5mL) of the pooled plasma were stored at -20°C until use. Fractions of 750 µL were collected in chilled deep-well microplates.

Recovery: In previous experiments, the entire UV absorbing material of the starting amount (~100%) was recovered with SEC [2, 4]. Currently we focus on proteins and thus, low molecular weight fragments and peptides eluting after collection (< 10 kDa) [7] were not considered in order to improve spreading and resolution of high molecular weight proteins.

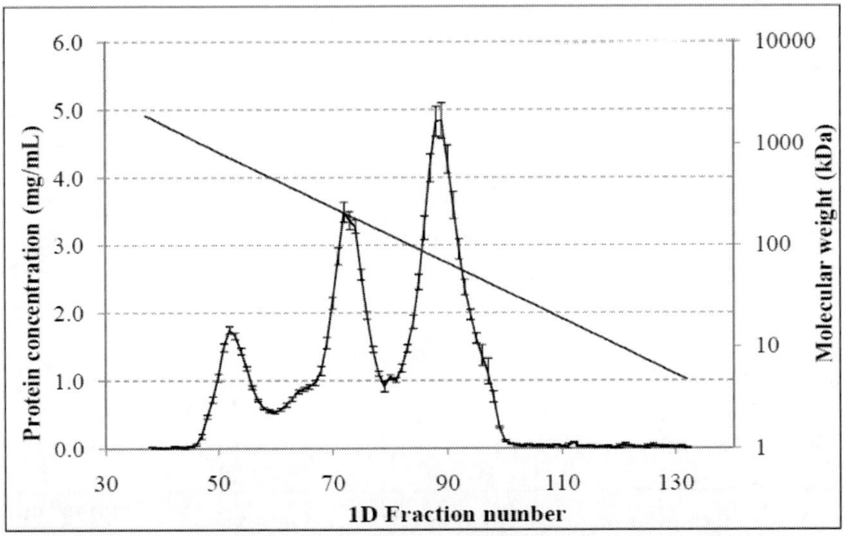

Means and SD (antennas) of protein concentrations of all 96 1D-fractions of one plasma sample are indicated (left ordinate). For comparison, straight line of column calibration is shown (right ordinate). Peaks containing characteristic proteins are indicated: lipoproteins (peak I), immunoglobulins (peak II) and albumin (peak III). Additionally, CV values are given in Tables 2 and 3.

Figure 19. Protein distribution and precision after 16 runs of 1D-SEC.

Following this, the total yield after 1D-fractionation was 93.73 ± 2.72 % (n=16) of the starting UV absorbing material and represent the entire protein content.

Precision: The standard deviation of 16 runs was determined for each 1D-fraction and is given as mean and CV (see Figure 19). Precision depends on the precision of readout and therefore on the level of protein concentration (Table 2), and location (Table 3). Best values are in the range of ~3%.

CV values are very high with 1D-fractions of low protein concentrations. These values are based on UV absorbances close to blanks and are known to be affected by high interference in microplates.

Moreover, 1D-fractions with the lowest protein content (0.02-0.04 mg/mL) and CV values above 50% yield 2D-fractions of very low protein concentration, since protein constituents are diluted on average to one third under 2D-AEC. Thus these fractions are not considered for proteomic searches by our method.

Table 2. Precision of 1D-separation

Protein concentration	Mean of CV	SD of CV	Number of 1D-fractions
3.0-5.0	4.50	0.89	9
1.0-2.95	4.51	2.04	27
0.3-0.97	5.19	5.70	15
0.05-0.18	21.15	12.76	6
0.02-0.04	51.44	25.84	21
0.02-5.0	3.22	0.22	18 best

CV values were sorted according total protein concentration (calculated from A_{280}-A_{700}, same data as in Figure 19).

Table 3. Dependence of CV values on elution volume (fraction number)

Peak	Fraction number	Mean of CV	SD of CV	Number of 1D fractions
I	49-54	3.68	0.36	6
II	66-77	3.83	0.41	6
III	79-83	3.74	0.75	5
III	93-96	7.45	3.59	4

1D-fractions with similar protein concentration, 1.1-1.9 mg/mL, were selected from all three elution peaks (same data as in Table 2, Figure 19).

6.2.2. 2D, Anion Exchange Chromatography

Recovery

Protein yield was determined from eight human, one bovine, one goat and two mice plasma samples, and two pooled human CSF samples. Protein content of all 2D-fractions were summed up, *i.e.,* the sum of up to 4128 fractions, and compared with the input, *i.e.,* the sum of the protein content of up to 96 1D-fractions. Thus, 97.1 ± 11.2% of the protein amount was recovered after 2D-fractionation, similar to previous data [2,16].

Precision

For the determination of precision 96 pools of all 96 1D-fractions of 16 1D-runs (see section 6.2.1.) were generated and 2D-separated with four and three column arrays per run, altogether 14 times. Various forms of precision parameters were determined after these 2D-runs.

Intra-Assay Precision

Since our method should be applied to 96 different 1D-fractions, knowledge of intra-assay precision determined with 96 identical samples subjected to a multitude of columns of one array is unfeasible. Rather, precision values were obtained with 96 parallel columns that separate 96 real samples, *i.e.,* 1D-fractions differing in protein compositions and total protein amounts and close to application conditions in proteomics. Thus, for determination of intra-assay precision, identical sets of aliquots of the pooled 96 1D-fractions were 2D-fractionated and the maximum number of replicates after one run applying the automated workstation was used, *i.e.* four arrays.

The means of CV values are better than ~10% with fractions that contain protein concentrations above 0.1 mg/mL, which represent ~ 15% of all fractions. Exemplarily results are shown in Figure 20.

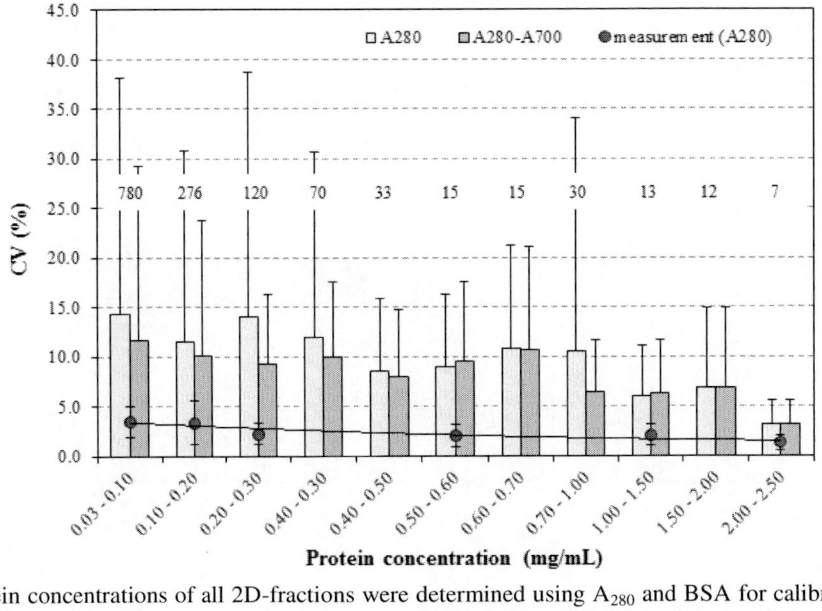

Protein concentrations of all 2D-fractions were determined using A_{280} and BSA for calibration. CV values were calculated on each 2D-location and sorted according to protein concentration, without (A_{280}) and with corrections by well-specific blank absorbance at 700 nm (A_{280}-A_{700}). Bars represent mean + SD of CV values within joined concentration intervals given on the abcissa. Numbers over the bars indicate number of locations of the respective group of 2D-fractions, circles indicate mean ± SD of the UV based protein determination method (n=8) without fractionation.

Figure 20. Intra-assay precision of 2D-separation.

Notwithstanding, a huge number of fractions exhibit protein concentrations of 0.03-0.1mg/mL and CV values up to ~40% obtained within series on one day. CV of fractions below 0.03 mg/mL is even higher, but these fractions were set to blanks and not further analyzed due to sensitivity requirements.

Inter-Assay Precision

A higher imprecision is found with runs on different days, even with identical column arrays (inter-assay, Figure 21). Thus, for reliable comparison, samples should be prepared on one day.

For comparison of more than four samples, these have to be separated in different series'. In this case, CV values were higher than within one series with most 2D-fractions viable for mass spectrometry. Mainly, CV values better than at least 50% may be guaranteed (Figure 21).

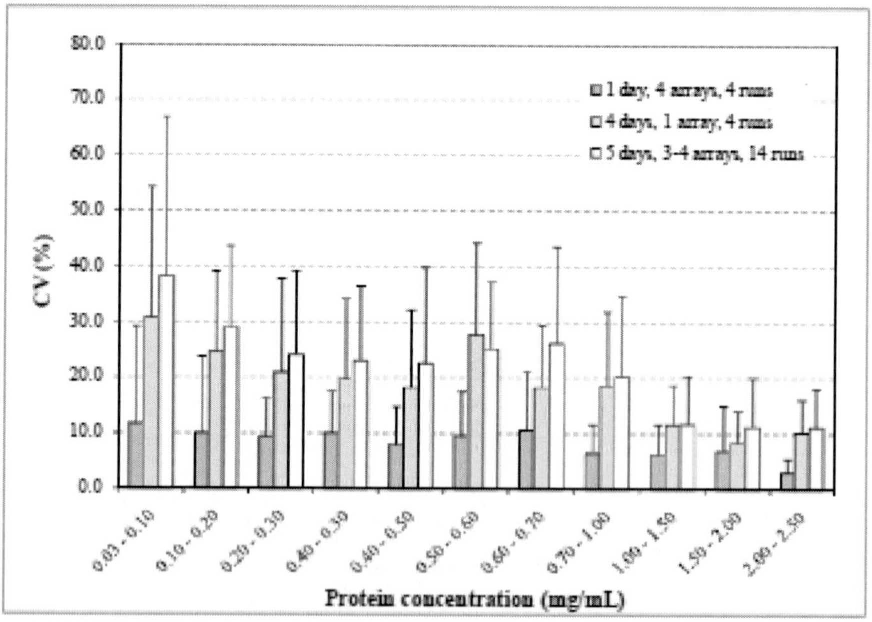

Intra-assay precision: 1 day, 4 arrays, four runs (same data as in Figure20). Inter-assay precision with one array: 4 days, 1 array, four runs. Inter-assay precision with several arrays: 5 days, 3-4 arrays per day, 14 runs.For details see Figure 20. Protein concentrations and CV values were determined on each 2D-location with correction by well-specific blank absorbance at 700 nm (A_{280}-A_{700}).

Figure 21. Comparison of various precision data obtained with 2D-separation.

Visual Quality Control

After 2D-separation, subsequent preparation and mass spectrometric analysis are tremendously laborious and time consuming. Since precision data have no further impact for individual ongoing runs, an online quality assessment is required to avoid depletion of resources.

For mammalian samples, an effective quality control by addition of microbial phycobilins was successfully evaluated.

These proteins carry chromophores that can be easily seen and read, show characteristic molecular weights and charges, elute highly reproducibly from AEC-columns, and do not disturb mass spectrometric analysis [15].

6.2.3. 3D, Lectin Affinity Chromatography

To determine binding capacity vast amounts of purified model proteins (transferrin and α_1-antitrypsin) were applied. The sugar specific eluate was defined as binding capacity. To test reproducibility of LAC columns with single lectins and lectin mixtures, the two purified proteins alone or in mixture (1/1, w/w)) were separated repeatedly. To this end, 10 µL binding buffer were spotted into the wells of a 384-well UV-Star microplate. Transferrin and α1-antitrypsin were used diluted to 0.56 mg/mL. 90 µL containing 50 µg were applied per run, *i.e.* ~1/8 of the maximum capacity (see below).

Elution and regeneration buffers were applied four times with corresponding volumes. Four fractions eluted by identical elution buffer were pooled to 3D-fraction I-IV.

a. Single Lectins, Single Proteins

Binding Capacity

The binding capacities of the Concanavalin A and WGA resins were determined with purified human transferrin and α_1-antitrypsin, respectively: 5.5 mg transferrin/g resin (440 µg/column) and 12.4 mg α_1-antitrypsin /g resin (992 µg/column).

Recovery and Precision

With 10 runs in series, 91.8 ± 4.6% of transferrin (intra-assay CV=5.01%) and with 10 different series′, 90.13 ± 3.2% transferrin (inter-assay CV=3.58%) were recovered by sugar specific elution (fraction II).

Additional fractions of the protein were recovered in I, III and IV to ~100% altogether.

b. Lectin Mixture, Mixtures of Two Purified Proteins

Intra-Assay Recovery and Precision

With mixtures of transferrin and α_1-antitrypsin in 14 runs on 14 columns within one series, 88.8 ± 3.6% of the applied protein amount were recovered. Precision of protein content within 3D-fractions were found with the following percentages (mean ± SD of the recovered protein amount): I: 11.2 ± 3.5 (CV= 31.9%), II: 57.1 ± 3.6 (CV=6.4%), III: 27.7 ± 1.4 (CV=5.1%), and IV: 3.8 ± 1.5 (CV=38.2%). Thus, precision depends on fractional protein amount analogous to 2D-fractions.

Inter-Assay Precision

After nine runs with single columns on 9 days, fraction I amounts to 15.0± 6.5% (CV= 43.5%), fraction II to 52.7 ± 5.2% (CV=9.9%), and fraction III to 28.8 ±2.1% (CV=7.3%).

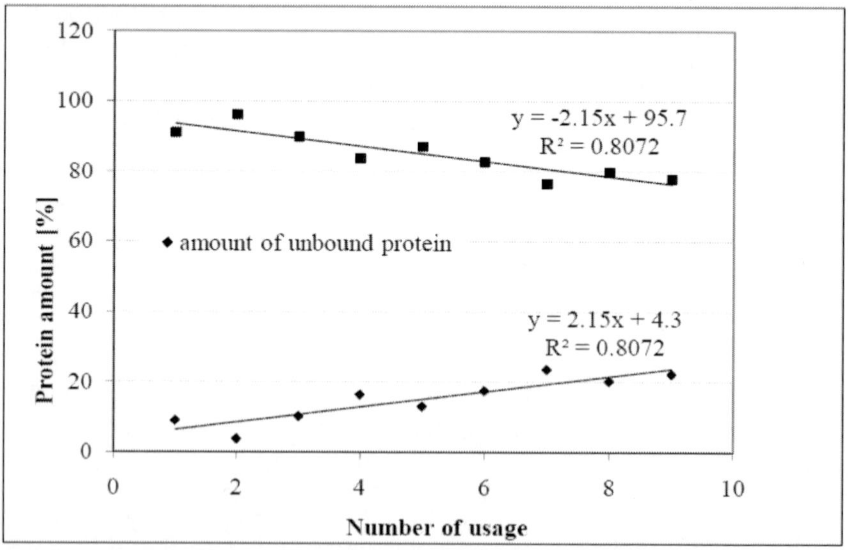

50 μg of a mixture of transferrin and α1-antitrypsin (1/1, w/w) was fractionated on LAC-columns (n=6) containing a mixture of ConA and WGA (1/1, v/v). Recovered protein amount was set to 100%. Unbound protein corresponds to fraction I.

Figure 22. Drifts of protein amounts within different 3D-fractions with repeated use.

Long Term Stability

Figure 22 indicates that there are small but significant drifts with repeated use. The percentage of fraction I, *i.e.* unbound flow-through rises, whereas protein content of the sum of fractions II - IV declines. Two factors may be responsible: denaturation of the ligands and/or escape of non-covalently bound subunits of lectinmultimers. The latter was qualitatively confirmed by mass spectrometry. This indicates that binding capacity decreases slightly by ~2% per run. Thus, the number of reuses is limited depending on the expected protein load and the comparative use in proteomics is feasible only within one series.

6.3. Precision of Mass Spectrometric Readout

From all 14 runs for evaluation of precision and recovery (see section 6.2.2., this chapter) six 2D-locations were selected from three different protein areas. These 6x14 fractions were individually prepared, digested, and individually analyzed by MS in duplicate.

Several parameters were analyzed quantitatively using features of Proteome Discoverer 1.3 and SIEVE 2.0.

6.3.1. Precision of Peak Intensities

Here, exemplary results with fraction 72_02 (1D_2D) of the immunoglobulin spot (cf. Figure 18) containing 2.2mg/mL and fraction 87_09 of the serum albumin area (1.06 mg/mL) are shown. As can be seen qualitatively in Figures 23 and 24, the chromatograms are highly reproducible.

After alignment of all chromatograms of the respective fraction alignment scores of 0.87 ± 0.028 (CV=3.16%) and 0.85 ± 0.049 (CV=5.79%) were obtained with fraction 72_02 and 87_09, respectively. For comparison, with identical samples used for quality control containing a tryptic digest from human transferrin (5 mg/mL) 0.93 ± 0.055 (n=4, CV=3.28%), the score is only slightly better. The results with all other tested fractions are similar.

6.3.2. Precision of Areas under the Curve

AUC of peaks were automatically selected and calculated by Thermo Xcalibur 2.1 (Qual Browser). Figure 25 shows peak selection and Figure 26 shows means and SD of selected peaks of low (1) and high (2) intensities out of two fractions.

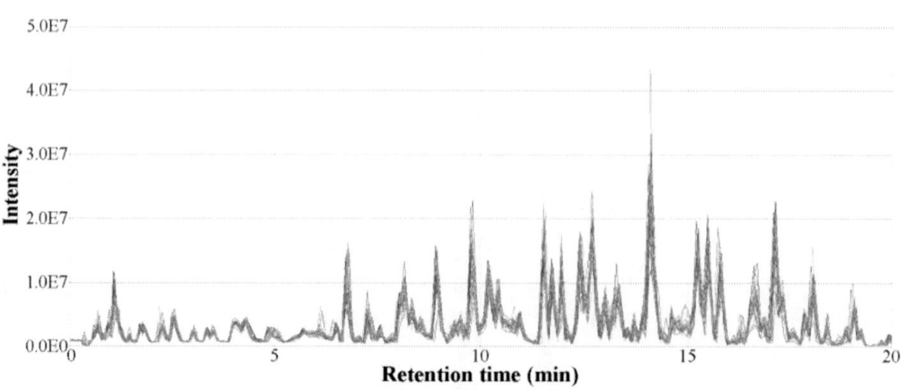

Alignment (SIEVE 2.0) of 26 individual LC runs of 13 2D-fractions at location 72_02 from 13 individual AEC runs (1 AEC run was omitted due to technical failure).

Figure 23. Overlay of peak intensities of all runs of 2D-location 72_02.

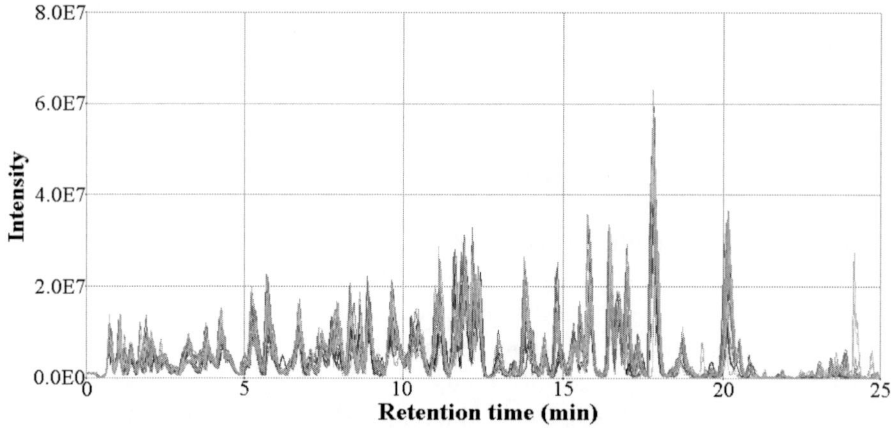

Alignment (SIEVE 2.0) of 28 individual LC runs of 14 2D-fraction at location 87_09 from 14 individual AEC runs.

Figure 24. Overlay of peak intensities of all runs of 2D-location 87_09.

CV values with low (1) and high (2) intensity were 24.30% and 15.48% with fraction 72_02, and 28.61% and 28.94% with fraction 87_09, respectively.

For comparison, with identical samples of a tryptic digest from human transferrin used for quality control (5 mg/mL, n=4, see section 2.7, this chapter) the precision is only slightly better with CV values of 6.35% and 21.42% for peaks with low and high intensity, respectively.

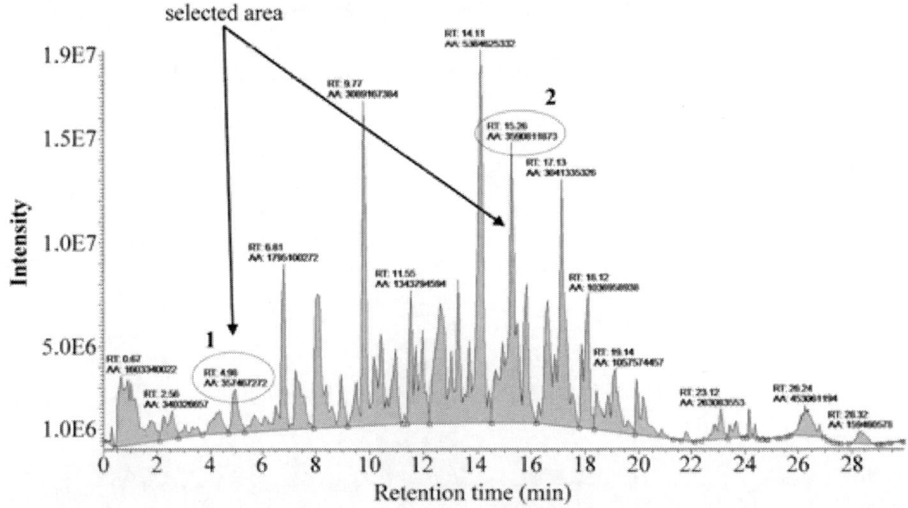

The figure shows one example (one run of location 72_02) of the automated determination of the intensity of selected peak areas by Thermo Xcalibur 2.1 (Qual Browser). Circles mark the selected regions for the calculation of AUC with low (1) and high (2) intensity at retention times of 4.98 and 15.26 min.

Figure 25. Intensity of ion current of one LC_MS run with location 72_02 with indication of selected peaks.

6.3.3. Reproducibility of Proteins Identification

Proteins were identified by the Proteome Discoverer 1.3 and Sequest database search algorithm applying a false discovery rate of <1%. As shown in Figure 27, similar numbers of identified proteins could be found in all replicates of the same fraction applying identical criteria. In total, with support by ≥2 peptides 16 (73.8% of the mean) and 5 (49% of the mean) proteins were found collectively in all samples of location 72_02 and 87_09, respectively.

Figs. 26 and 27 would be better located before section 7.

7. Applications

7.1. Profiling of Proteomes of Various Sample Types

Besides recovery and precision, the final yield of identified components is one of the quality parameters regarding comprehensiveness and sensitivity of a proteomic workflow. All these parameters combined account for success in biomarker search.

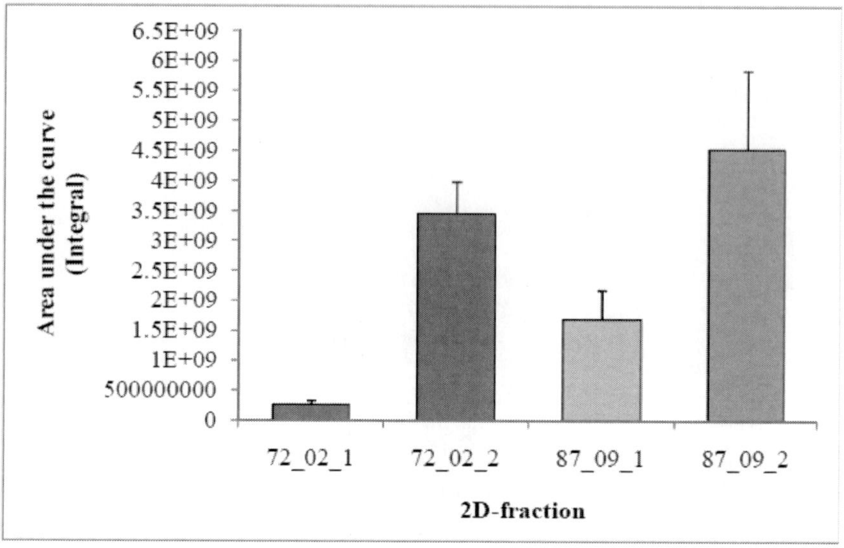

From two 2D-locations (72_02, 87_09) two peaks were selected (see Figure 25) with different intensities and different retention times. Bars represent means + SD of AUC of selected peak areas for all runs with two 2D-locations as indicated.

Figure 26. Mean and SD of four areas under the curve of the intensity of ion current of LC_MS runs with two 2D-locations.

Thus, sample input amount, sample separation, and mass spectrometric readout including database size contribute equally and critically to the result.

The previously published semiautomatic sample preparation of two human serum samples yields 2669 and 520 identified non-redundant proteins supported by ≥1 peptide and by ≥2peptides, respectively [16]. The previous procedure differs only slightly from the current one with serum instead of plasma samples and AEC on DEAE-cellulose instead on DEAE-toyopearls. Moreover, readout was previously performed by nLC-MS using an Ultimate system (Dionex) and a Qstar pulsar (Applied Biosystems GmbH, Darmstadt, Germany) instead of μUHPLC combined with a LTQ Orbitrap Discovery equipped with an H-ESI source. The current equipment allows about a six fold throughput (see section 2.6, this chapter) compared to the former one.

Hence, a slightly higher number of 3166 and 816 non-redundant proteins could be identified supported by ≥ 1 and ≥ 2 peptides, respectively (cf. Figure 28). This yield was obtained by applying all presented tools and devises on every second 2D-fraction (with >0.03 mg/mL) of two human plasma samples and analysis of technical duplicates. Surprisingly, only ~25% of the identified hits (supported by ≥ 2 peptides) represent collective proteins.

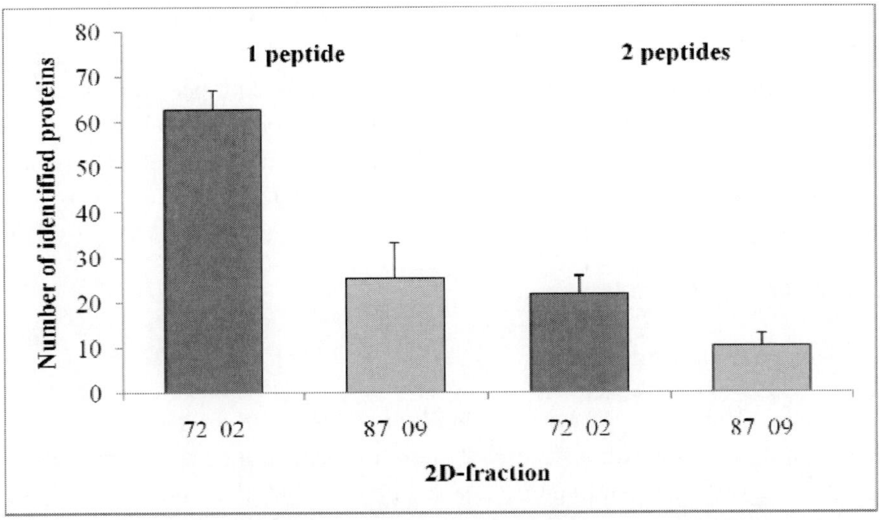

Numbers of identified proteins were determined using the Proteome Discoverer 1.3 at location 72_02 (n=26) and 87_09 (n=28) from 13 and 14 individual 2D-runs, respectively. Bars represent the means + SD of identified proteins supported by ≥1 peptide (left parts, mean 62.6 (72_02) and 25.3 (87_09)) and supported by ≥2 peptides (right parts, mean 21.7 (72_02) and 10.2 (87_09)).

Figure 27. Precision of the number of identified protein hits.

Bright: hits found only in sample 1, grey: hits found only in sample 2, dark grey: overlap (collective proteins); left part: identified by ≥1 peptide, right part: identified by ≥2 peptides; areas correspond to hit numbers.

Figure 28. Number of identified proteins from two different individual samples from two healthy volunteers of EDTA-plasma.

Additionally, our recent method could be applied easily to bovine and goat plasma with similar protein input compared to human samples, to pooled and concentrated human cerebrospinal fluid with about 1/4, and to mouse serum with about 1/10 protein input compared to human plasma samples, respectively.

So, with one bovine and one goat plasma sample, with two pooled human cerebrospinal fluids, and with one mouse serum sample 1151, 1044, 1368, and 378 non-redundant proteins could be identified with ≥1peptide hit per protein. One has to note that the actual (07/2013) number of data base entries in UniProtKB with ruminants (~7.700) is much lower than for humans (~812.000) and mice (~80.000), and that the protein amount of mouse serum is one order lower.

7.2. Biomarker Search

7.2.1. Severe Inflammation

Sepsis, severe sepsis, and septic shock are states with increasing severity of inflammation. Sepsis is still a signifcant cause of death at intensive care units. For diagnosis, various vital parameters, laboratory and hemodynamic data, and data reflecting organ function are used. Parameters used for diagnosis are related to pathogenesis involved, *i.e.*, oxidative stress, apoptosis, dysregulation of innate immunity, endothelial dysfunction, coagulation and complement activation, metabolism, and signal cascades. However, these parameters cannot diagnose sepsis reliably nor can they predict progression or prognosis. Hence, the search for useful proteomic biomarkers is urgent and still ongoing [*e.g.*, 17,18,19,20].

Our method under development yielded 36 biomarker candidates in total. Out of these, CRISP and A1B-GP and a seemingly low-molecular weight albumin fraction were evaluated and could be validated by immune assays [5].

7.2.2. Nephropathies: Alport Syndrome and End Stage Renal Disease

The Alport syndrome is a hereditary clinically and genetically variable nephropathy which develops into nephrotic syndrome and finally, End-stage renal disease (ESRD). Additionally, extra-renal manifestations may occur, *e.g.*, sensory-neural deafness, ocular abnormalities, and leiomyomatosis. Although inherited, therapeutic intervention in renal failure may be achieved by several principles. Since therapy by Ramipril is particularly effective when started in early or even preclinical stages [21], early diagnosis is urgently needed. However, reliable non-invasive markers for screening, diagnosis and therapy monitoring are lacking.

Until now, reliable diagnosis requires biopsy and/or genetic methods. Generally, these procedures are not performed before the onset of strong clinical symptoms.

Therefore we applied our method to biomarker search in Alport syndrome. Thereafter, some of our biomarker candidates identified in serum [5] were evaluated in a cohort of patients exhibiting Alport syndrome, other nephropathies, and healthy controls. Out of them, a panel of parameter combinations was found to discriminate between patient groups and thus verifying our biomarker candidates [22, 23].The same method was applied to patients with ESRD of diverse origin undergoing hemodialysis. Here, surprisingly, long-term serum proteomes were found to be quite similar under treatment with two variants of hemodialysis (*i.e.*, by high- and low-flux membranes). Contrary to this, 29 significantly altered serum protein constituents were identified in ESRD patients in comparison to healthy controls [8].

7.2.3. Psoriasis

Psoriasis vulgaris is a frequent polygenetic inherited and chronic inflammatory disease with the primary focus on skin. Psoriasis affects 2-4% of the population worldwide. A large proportion of patients unpredictably develop associated diseases as *e.g.*, atherosclerosis, myocardial infarction, uveiitis, Mb. Crohn, and various metabolic und psychiatric alterations [24]. In 20-30% of cases the disease is associated with Psoriatic arthritis (PsA) [25]. Currently, the diagnosis of Psoriasis and PsA is mainly based on clinical assessment and there are no diagnostic tests or biomarkers with high sensitivity and specificity, especially in the early disease state [26]. Thus, the current method was applied to EDTA-plasma of patients with Psoriasis, Psoriasis arthritis and controls. The study is still in progress. Some preliminary results are shown in Table 4 (A-C):

Table 4. Preliminary results with Psoriasis with one sample couple (patient/control)

A: Global results

Parameter	Criteria	Number
total analyzable 2D fractions	>0.03 mg/mL in at least one sample	1053
fractions with protein conc↑	UV absorbance, factor ≥ 4	31
fractions with protein conc↓	UV absorbance, factor ≤ 0.25	17
analyzed fractions	~every fourth of both samples	530
peptide hits ↑	Total, FDR <0.01, ratio ≥ 4	634
peptide hits ↓	Total, FDR <0.01, ratio ≤ 0.25	1338

B: Identification results

	Criteria	Condensed number of protein hits	
	According Expasy entries→	Biomarker candidates	typical plasma proteins
protein hits ↑	identified ≥twice with alike direction in joined fractions	29	36
protein hits ↓		20	62

C: Examples of biomarker candidates

ID	Name	Biological function
tr|H0YAS8	Clusterin alpha chain	not known
tr|B7Z8R6	Trypstatin, Alpha-1-microglobulin	not known
sp|A8MT79	Putative zinc-alpha-2-glycoprotein-like 1	not known
sp|P04264	Keratin type II cytoskeletal 1	complement activation epidermis development, carbohydrate binding

EDTA-Plasma of Patients with Psoriasis and age and gender matched controls were processed in parallel including mass spectrometry. UV absorbance and mass spectrometric data of homologous fractions were compared.

7.2.4. Amyotrophic Lateral Sclerosis

ALS is a progressive neurodegenerative disease of upper and lower motor neurons with systemic implications. About 10% of ALS patients indicate a positive family history (familial ALS). The remaining 90% are so-called sporadic ALS cases [27]. ALS is hallmarked by central paresis, spastic tonicity, uninhibited muscle reflexes, fasciculation, pyramidal signs and muscle atrophy (amyotrophia) [28]. Most of the patients perish approximately 2-5 years after disease onset by respiratory paralysis. There is no cure for this rapidly disabling disorder. The only available therapy is based on Riluzole. This drug extends the life expectancy for merely 2-3 months [29]. As yet, it is still unclear which cellular events initiate the degeneration of motor neurons including corticospinal tract, brainstem, and spinal anterior horn neurons. As a consequence, the search for ALS biomarker candidates is urgently needed.

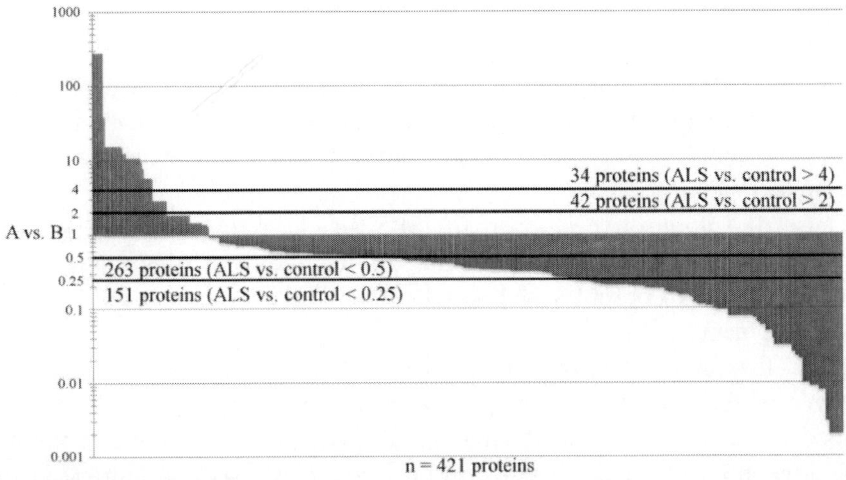

Each vertical line represents one hit with SIEVE 2.0. Horizontal lines indicate borders at a two- and fourfold change under disease.

Figure 29. Ratios of intensities of identified proteins in CSF from patients with ALS (A) and controls (B).

Reliable biomarkers have not only the potential to provide early diagnostic tools, they could further provide molecular indicators that help to gain deeper insights into the pathophysiology of ALS.

This is constitutive for the discovery and development of therapeutic interventions by determining whether a potential drug candidate targets the disease in an effective manner [30].

Therefore we began to apply our method to biomarker search in ALS. CSF samples were obtained after written declaration of consent from patients undergoing diagnostic lumbar puncture who agreed to give extra CSF for proteomic analysis by strict standard operation procedure.

Prior to automated sample preparation, matched CSF samples of ALS patients (n=7) and controls (n=9) have been pooled and subsequently concentrated by filtration through a spin column (Vivaspin 500, Sartorius, MWCO 5 kDa) in order to get enough protein input.

Although still in progress, as an initial result, several hundred protein species can be shown to be stringently different in ALS (Figure 29), and require further evaluation.

Conclusion

The presented method was compiled in strict compatibility with microplate technology throughout: from separation, readout and quantitation, sample preparation between separation steps, storage, up to stacking within the autosampler of LC-MS.

Thus, all equipment developed for SBS standards may be used including liquid handling techniques and automation facilities. Existing gaps of still unavailable tools, disposables and software were filled prototypically with our own developments.

Thus, maximum standardization is achieved by true parallel processing. Moreover, with the majority of sample preparation steps the parameters, precision and recovery can be determined using spectrophotometric readout in contrast to other proteomic methods. These parameters were used for strict process optimization. Although very precise, using our automated platform CV values and recoveries of singular steps could never achieve 0% and 100%, respectively. Thus, there is some imprecision to be considered with the final results as for other analytical tools. Currently, we are not able to determine overall precision experimentally. Furthermore, even if measureable, this statistical data is without impact for the interpretation of a distinct experiment. Since the majority of 2D-fractions achieve CV values of lower than 50% for both total protein concentrations after fractionation and exemplarily with mass spectrometric readout, a fourfold change of the final result seems to be an acceptable lower boundary for defining significance in biomarker search.

Although entire sample constituents are preserved and prepared for analytics, sample preparation became fast and highly reproducible. Now, mass spectrometry was identified as bottleneck of throughput.

Due to comprehensiveness and known low imprecision, our platform offers a valuable alternative, complementary to others, for proteomic studies and biomarker search driven without hypotheses.

Acknowledgments

The editorial assistance of Laura McMillan, the financial support of BMBF (PTJ-Bio/0312699), DFG (Ho 1311/3–1), the Thuringia Ministry for Education, Science and Culture, TAB and the EFRE-fund (2010 FE 9001), the BMWi (03EGSTH003, KF2937601FR1), the IZKF (O. Witte), the Dermatology (J. Norgauer), the Institute of Biochemistry I (B. Qualmann), all University Hospital

Jena, the technical support of the Central Scientific Workshop of the University Hospital, and the supply of plasma and CSF samples by the Departments of Dermatology and Neurology of the University Hospital, and the Friedrich Löffler Institute Jena is gratefully acknowledged.

References

[1] Krüger, T., Lehmann, T., Rhode, H. Effect of quality characteristics of single sample preparation steps in the precision and coverage of proteomic studies-A review. *Anal. Chim. Acta*. 2013 May 7;776:1-10. PubMed PMID: 23601274. Epub. 2013/04/23. eng.

[2] Horn, A., Kreusch, S., Bublitz, R., Hoppe, H., Cumme, G. A., Schulze, M., et al. Multidimensional proteomics of human serum using parallel chromatography of native constituents and microplate technology. *Proteomics*. 2006 Jan.; 6(2):559-70. PubMed PMID: WOS: 000235207700018.

[3] Bublitz, R., Kreusch, S., Ditze, G., Schulze, M., Cumme, G. A., Fischer, C., et al. Robust protein quantitation in chromatographic fractions using MALDI-MS of tryptic peptides. *Proteomics*. 2006 Jul.;6(13):3909-17. PubMed PMID: WOS:000239045500018.

[4] Kreusch, S., Schulze, M., Cumme, G. A., Ditze, G., Moore, T., Rhode, H. Searching biomarker candidates in serum using multidimensional native chromatography. I. Enhanced separation method. *Journal of chromatography B, Analytical technologies in the biomedical and life sciences*. 2008 Nov. 15;875(2):567-72. PubMed PMID: 18952511. Epub. 2008/10/28. eng.

[5] Baum, A., Pohl, M., Kreusch, S., Cumme, G. A., Ditze, G., Misselwitz, J., et al. Searching biomarker candidates in serum using multidimensional native chromatography. II Method evaluation with Alport syndrome and severe inflammation. *Journal of chromatography B, Analytical technologies in the biomedical and life sciences*. 2008 Dec. 1;876(1):31-40. PubMed PMID: 18952508. Epub. 2008/10/28. eng.

[6] Cumme, G. A., Kreusch, S., Nagel, M., Rhode, H. Multidimensional chromatography: validation and efficient fishing for biomarkers and fractions containing them using the VisualCockpit software package. *Proteomics*. 2008 Jan.;8(1):37-41. PubMed PMID: 18095362. Epub. 2007/12/21. eng.

[7] Pesek, J., Krüger, T., Krieg, N., Schiel, M., Norgauer, J., Grosskreutz, J., et al. Native chromatographic sample preparation of serum, plasma and cerebrospinal fluid does not comprise a risk for proteolytic biomarker loss. *Journal of chromatography B, Analytical technologies in the biomedical and life sciences*. 2013 Apr. 1;923-924:102-9. PubMed PMID: 23500354.

[8] Hallbauer, J., Kreusch, S., Klemm, A., Wolf, G., Rhode, H. Long-term serum proteomes are quite similar under high- and low-flux hemodialysis treatment. *Proteomics Clinical applications*. 2010 Dec.;4(12):953-61. PubMed PMID: 21137035.

[9] Wen, Z., Niemeyer, B. Evaluation of two different Concanavalin A affinity adsorbents for the adsorption of glucose oxidase. *Journal of chromatography B, Analytical technologies in the biomedical and life sciences.* 2007 Sep. 15;857(1):149-57. PubMed PMID: 17683998.

[10] Kreusch, S., Schwedler, S., Tautkus, B., Cumme, G. A., Horn, A. UV measurements in microplates suitable for high-throughput protein determination. *Anal. Biochem.* 2003 Feb. 15;313(2):208-15. PubMed PMID: 12605857. Epub. 2003/02/28. eng.

[11] Maischak, H., Tautkus, B., Kreusch, S., Rhode, H. Proteomic sample preparation by microdialysis: easy, speedy, and nonselective. *Analytical biochemistry.* 2012 May 15;424(2):184-6. PubMed PMID: 22342948. Epub. 2012/02/22. eng.

[12] Kraus, H. PVC-Separatoren als Elektroden-Trennelemente in Akkumulatoren. *Plasteverarbeiter*. 1976; 5:1-6..

[13] Rhode, H., Kreusch, S., Endmann, H., Händel, M., Sammler, G., Zimmermann, J., *Inventors Device for resaving, treating and storing small volume samples2008*.

[14] Maischak, H., Tautkus, B., Kreusch, S., Rhode, H. Proteomic sample preparation by microdialysis: Easy, speedy, and nonselective. *Analytical Biochemistry.* 2012;424(2):184-6.

[15] Pesek, J., Krüger, T., Tautkus, B., Rhode, H.Colorful quality control of chromatographic sample preparation. *J. Chromatogr*. B. 2013, , in press, (DOI) 10.1016/j.jchromb.2013.07.001

[16] Kreusch, S., Schulze, M., Cumme, G. A., Ditze, G., Moore, T., Rhode, H. Searching biomarker candidates in serum using multidimensional native chromatography. I. Enhanced separation method. *J. Chromatogr*. B. 2008 Nov.;875(2):567-72. PubMed PMID: WOS:000261272900034.

[17] Siqueira-Batista, R., Mendonca, E. G., Gomes, A. P., Vitorino, R. R., Miyadahira, R., Alvarez-Perez, M. C., et al. Proteomic updates on sepsis.

Revista da Associacao Medica Brasileira. 2012 Jun.;58(3):376-82. PubMed PMID: 22735232.

[18] Vincent, J. L., Beumier, M. Diagnostic and prognostic markers in sepsis. *Expert review of anti-infective therapy*. 2013 Mar.;11(3):265-75. PubMed PMID: 23458767.

[19] Singer, M. Biomarkers in sepsis. *Current opinion in pulmonary medicine*. 2013 May;19(3):305-9. PubMed PMID: 23411577.

[20] Wacker, C., Prkno, A., Brunkhorst, F. M., Schlattmann, P. Procalcitonin as a diagnostic marker for sepsis: a systematic review and meta-analysis. *The Lancet infectious diseases*. 2013 May;13(5):426-35. PubMed PMID: 23375419.

[21] Kashtan, C. E., Ding, J., Gregory, M., Gross, O., Heidet, L., Knebelmann, B., et al. Clinical practice recommendations for the treatment of Alport syndrome: a statement of the Alport Syndrome Research Collaborative. *Pediatric nephrology*. 2013 Jan.;28(1):5-11. PubMed PMID: 22461141. Pubmed Central PMCID: 3505543.

[22] Pohl, M., Kreusch, S., Cumme, G., Ditze, G., Misselwitz, J., John, U., Rhode, H.Proteomics-based search for biomarkers in sera of patients with Alport syndrome. *Ped. nephrology*. 2009, 24:903.

[23] M. Pohl, K. Danz, O. Gross, U. John, J. Urban, L. Patzer, S. Habbig, M. Feldkötter, O. Witzke, M. Walther, H. Rhode. Diagnosis of Alport syndrome – search for proteomic biomarkers in body fluids. *Ped. nephrology,* 2013(DOI) 10.1007/s00467-013-2533-5, in press.

[24] Guttman-Yassky, E., Nograles, K. E., Krueger, J. G. Contrasting pathogenesis of atopic dermatitis and psoriasis--part I: clinical and pathologic concepts. *The Journal of allergy and clinical immunology*. 2011 May;127(5):1110-8. PubMed PMID: 21388665.

[25] Zachariae, H. Prevalence of joint disease in patients with psoriasis: implications for therapy. *American journal of clinical dermatology*. 2003;4(7):441-7. PubMed PMID: 12814334.

[26] Villanova, F., Di Meglio, P., Nestle, F. O. Biomarkers in psoriasis and psoriatic arthritis. *Ann. Rheum. Dis*. 2013 Apr.;72 Suppl. 2:ii104-10. PubMed PMID: 23532439.

[27] Li, T. M., Alberman, E., Swash, M. Comparison of sporadic and familial disease amongst 580 cases of motor neuron disease. *Journal of neurology, neurosurgery, and psychiatry*. 1988 Jun.;51(6):778-84. PubMed PMID: 3404186. Pubmed Central PMCID: 1033147.

[28] Chio, A., Calvo, A., Moglia, C., Mazzini, L., Mora, G., group Ps. Phenotypic heterogeneity of amyotrophic lateral sclerosis: a population

based study. *Journal of neurology, neurosurgery, and psychiatry*. 2011 Jul.;82(7):740-6. PubMed PMID: 21402743.

[29] Miller, R. G., Mitchell, J. D., Moore, D. H. Riluzole for amyotrophic lateral sclerosis (ALS)/motor neuron disease (MND). *Cochrane database of systematic reviews*. 2012;3:CD001447. PubMed PMID: 22419278.

[30] Kruger, T., Lautenschlager, J., Grosskreutz, J., Rhode, H. Proteome analysis of body fluids for amyotrophic lateral sclerosis biomarker discovery. *Proteomics Clinical applications*. 2013 Jan.;7(1-2):123-35. PubMed PMID: 23129563.

In: Automation Systems of the 21st Century
Editors: D. Arent and M. Freebush
ISBN: 978-1-62948-262-0

Chapter 2

MODERN AUTOMATION IN WAREHOUSING BY USING THE SHUTTLE-BASED TECHNOLOGY

Tone Lerher*

Faculty of Mechanical Engineering, University of Maribor, Maribor, Slovenia

Abstract

In this chapter, a conceptual model for calculating the system performance of shuttle based automated warehouse system (SBAWS) is presented. According to the global trends in material handling and warehousing, a SBAWS should encompass not only the installation of equipment with fastest drives, but also consider the energy consumption and environment effects on the installed equipment. The proposed conceptual model illustrates energy regeneration and throughput performance calculations of a SBAWS. Thus, the proposed model enables design of a high-throughput-performance SBAWS with reduced energy consumption through energy regeneration and so reduced CO2 emission. It is believed that by this study, through considering the throughput performance and the energy efficiency (regeneration) simultaneously a SBAWS will provide a significant contribution to the environment and the economy.

* E-mail address: tone.lerher@um.si; Tel.: 00 386 2 220 7723; Fax: 00 386 2 220 7994 (Corresponding author)

Keywords: warehouses, shuttle based technology, throughput performance, energy recovery, performance analysis

1. Introduction

Automation in material handling and warehousing processes has gained more significance recently. For instance, Material Handling Industry of America grew by around 12 % in the year of 2012 confirming that something has changed in the field of automation in material handling. The reason for the relatively high interest in automation is due to the fact that automated systems provide high throughput efficiency to handle split case order fulfillment, relatively fast ROI and most of the case go hand-to-hand with green logistics and ergonomic at work. The automation trend should remain to be the golden standard for automated material handling in warehousing processes and in the near future.

In this chapter, relatively a new technology in automated warehouse system - SBAWS - is introduced and evaluated. Today's warehouses and distribution centers usually operate on bases of the "person-to-goods" system. In the case of an order picking warehouse, once the order list is released, the order picker (warehouse worker) travels in the warehouse to retrieve the items from their locations based on the order list. The retrieved items are then packed and prepared to be shipped to customers. This order picking process usually requires inefficient long travel times within the warehouse. This travel time may be up to 60% of the total cycle time and this rate may change depending on several factors, such as layout of the warehouse, storage policy, order size, routing policy, etc.

Order picking warehouses have been successfully studied by many authors from the material handling and warehousing research community mostly with the objective function minimizing the travel time of the order picker. It must be emphasized, that the order fulfillment in the past usually represented the interaction between warehouses and wholesalers. Today there are so many warehouses and distribution centers supplying multi-channel order fulfillment (online shopping, smart phones shopping, etc.). This requires more frequent and relatively small orders to be shipped directly to final customers. Therefore, a traditional warehouse applying "person-to-goods" system would not be an effective way in today's market. One solution might be the application of relatively a new trend in "goods-to-person" order picking system supported with the shuttle-based technology.

Automation in warehousing processes first appears in 1960 with the application of automated storage and retrieval systems (AS/RS) for carrying unit-loads. The first 20 meters AS/RS was built in Germany in 1962 (Kartnig at al., 2012). A unit load consists of items in cartons which are stuck in a single pallet. In the 1990s next evolution of automated storage represents AS/RS for smaller and lighter loads in totes, cartons and bins, known as the mini-load AS/RS. The mini-load AS/RS use the storage and retrieval (S/R) machine with the hoisted carriage that can manipulate loads independently horizontally and vertically. The S/R machine can be either single aisle (aisle captive) or can be transferred through a transfer shuttle to the adjacent aisle. The benefit of the mini-load AS/RS is moderate investment, according to the throughput performance. Weaknesses of the mini-load AS/RS are high power consumption of the storage and retrieval machine and subsequent extensions of this system is somehow limited.

In order to achieve high throughput rate with high energy efficiency shuttle-based technology has been introduced to the world market by several system integrators like GEBHARDT, TGW and others. A SBAWS is composed of elevator having lifting table that is feeding the buffer position with totes (containers) at different levels of the storage rack. In each level there is a shuttle carrier that transports containers to their destination addresses within that level. This relatively new technology can provide high performance (can handle up to 1000 containers/hour) and has a flexibility to increase the number of shuttle carriers at any time. Moreover, this system is relatively energy efficient compared to mini-load AS/RS. However, this system may require high investment cost and highly complex management system for operating the shuttles (Kartnig at al., 2012). A shuttle based system is somehow competitive to a mini-load AS/RS and retrieval system only when there is a large throughput rate requirement where each shuttle's payload does not exceed 50 kg/shuttle (TGW and GEBHARDT).

In the next section, SBAWS will be described and discussed.

2. Shuttle Based Automated Warehouse System

As mentioned previously, a SBAWS is composed of elevator with lifting tables that are attached on a mast, shuttle carriers, buffer positions and the storage racks (see Figure 1).

The elevator through the lifting table moves containers up and down to the prescribed level in the storage rack. To be able to process more work two independent lifting tables may be attached to the elevator one of which is on the right side and the other one is on the left side of the mast. In this case, the performance of the elevator can be doubled. Another way to increase the performance of the elevator may be application of a special lifting table that can handle multiple containers at a time and therefore work on a basis of multi-cycle. An elevator can move lifting tables up to v_y = 1.5 m/s theoretically. An energy regeneration system module can be applied during lowering (braking) the lifting table. Elevators are usually bottleneck in this system so they usually determine the performance of the whole system.

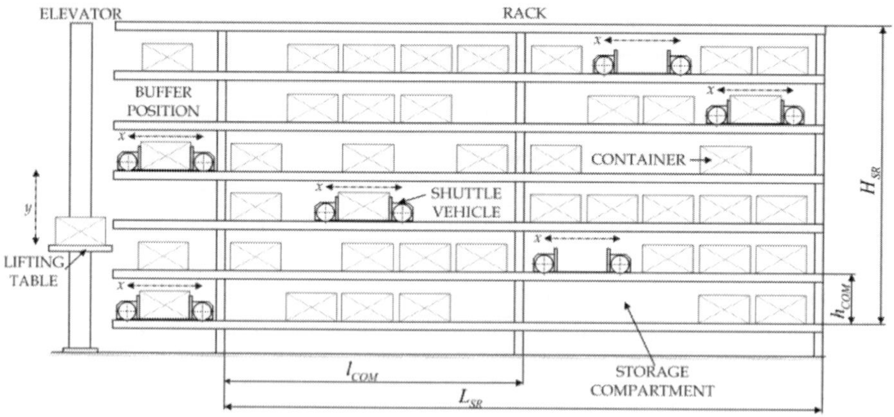

Figure 1. Shuttle based automated warehouse system.

A shuttle carrier is a small autonomous vehicle with four wheels that transports containers from the buffer position to the storage locations in the storage racks. It is equipped with telescopic attachment for manipulating containers. The maximal weight of a container should not exceed 50 kg/shuttle carrier and its dimensions should be in the range of: min. (150 x 200 x 80) mm, max. (600 x 400 x 250) mm. A shuttle carrier can travel up to v_x = 4 m/s theoretically. An energy regeneration system module can also be applied during its braking. Usually there is a single shuttle carrier in each level of a storage rack. This assumption can be released if we use a special shuttle elevator at the back of the storage rack, for moving shuttle carriers up and down to the prescribed level in the storage rack.

The buffer position is placed at each level of the storage rack and is used for buffering the containers from the elevator and shuttle carriers.

The storage rack is composed of storage compartments that can receive n loads. By multiplying storage compartments in the horizontal and in the vertical direction, the length and the height of the storage rack is achieved. The storage rack can be implemented as a single or a double deep.

One of the important decision problem in a SBAWS is how to calculate the system performance (throughput capacity) of the elevator and the shuttle carrier. In the literature, many analytical, heuristic and numerical models exist to calculate the system performance of AS/RS. A comprehensive explanation of the current state of the art of AS/RS has been presented by Roodbergen and Vis (2009). On the other hand there is a lack of research for SBAWS. Malmborg (2002) has presented autonomous vehicle storage and retrieval systems (AVS/RSs), which are characterized by horizontally operating vehicles sharing a fixed number of lifts for vertical movement. Ekren at al. (2010) have presented simulation based experimental design to identify factors affecting performance of AVS/RS. Four factors - dwell point policy, scheduling rule, I/O locations, interleaving rule - are defined and their interaction effects are identified on three performance measures via the design of experiment. Ekren and Heragu (2012) study performance comparison of two material handling systems - AVS/RS and traditional crane-based AS/RS (CBAS/RS). They consider aisle-to-aisle CBAS/RS and several performance measures in the comparison procedure. They also calculate investment costs of the two systems based on several designs. Although the investment cost of an AVS/RS is higher than the studied CBAS/RS, the other performance measures are better than the CBAS/RS. Hector and Vis (2012) have presented sequencing model for dynamic storage systems with multiple lifts and shuttles. The study with the scheduling problem of two lifts has been analyzed. An integrated look-ahead strategy heuristic to simultaneously assign a set of pre-defined requests to the lifts and the order in which they will be handled has been proposed by the authors.

Based on the literature review, none of the presented models deals with analytical or numerical models for calculating system performance along with the energy efficiency (energy regeneration) of the system. So far, the existing studies mostly have dealt with the objective function minimizing travel time and total cost. To the best of my knowledge, in the literature there is no study taking into consideration energy efficiency in the system's design. In this chapter, it is also aimed to complete an energy regeneration study for a SBAWS to consider energy efficiency in the design concept.

It must be emphasized that the European Union legalized regulations like Directive of the European Parliament and of the Council 2006/32/WE from 5thApril, 2006 on energy end-use efficiency and energy services and repealing

Council Directive 93/76/EEC. Also the Kyoto protocol somehow commands the signed countries to lower the CO_2 emissions for 20 % until the year 2020.

According to global trends and recent development in green technologies, there is a necessity to take into account the energy efficiency along with the usually applied objectives (minimum travel time and maximum throughput). In the following sections, a model for calculating the system performance ofa SBAWS is presented and evaluated.

3. A Model for Calculating the System Performance of a SBAWS

The model for calculating the system performance of a SBAWS is based on the structured approach, where all parameters influencing the warehouse performance are taken into account. The model includes the following main modules.

Definition of the Throughput Performance of the Elevator

As already mentioned, the elevator with lifting tables (one lifting table is attached on the right and the other one is attached on left side of the mast) has important influence on the system performance. In practice, the warehouse integrators usually operate with data that the system (elevator) could manipulate up to 1000 containers/hour. Since there are no known "basic" analytical or numerical models to prove the performance of the elevator, a conceptual analytical travel time model is presented. Generally, the throughput of an elevator can be increased by using efficient drives with high vertical velocity profile, with the implementation of lifting table that can handle up to three containers at a time (multi-cycle) and with the implementation of new heuristics to schedule lifting tables.

In the present study, the lifting table can carry one container at a time. The conceptual analytical travel time model of the elevator is developed by considering single and dual command cycles with random assignment policy. For the proposed conceptual analytical travel time model of the elevator, the analytical model of the high performance transfer car presented by Oser and Reisinger (2008) has been used.

Definition of the Throughput Performance of a Shuttle Carrier

A shuttle carrier operates in the prescribed level of an aisle and carries loads from the buffer position to the storage locations in the storage racks and vice versa. Since lifts are usually bottlenecks in this system, allowing level-to-level shuttle vehicles in other words, shuttle vehicles that can travel between different levels of an aisle may not change the average cycle time of a load in the system. Moreover, this may decrease the initial investment cost of the system.

Since there is no "basic" analytical or numerical model to calculate the throughput performance of a SBAWS, a conceptual analytical travel time model is presented by this study. Generally, the throughput performance of the shuttle carrier can be increased by using the efficient drives with high horizontal velocity profile, by using the implementation of double deep storage system and with the implementation of new heuristics to schedule the shuttle carriers.

In the present study, the shuttle carrier can receive one container at a time in a single deep storage rack. The conceptual analytical travel time model of the shuttle carrier is used for the operation of single and dual command cycles with random assignment policy. For the proposed conceptual analytical travel time model of the shuttle carrier, the modified FEM 9.851 is used (FEM, 2001).

Definition of the Energy Regeneration of the Elevator and the Shuttle Carrier

Due to the limited energy resources in the world and importance of green environment concept, energy consumption (energy regeneration) and the concept of "zero CO_2 emission" issues have become significant issues recently. It is generally known that the distribution centers (buildings), along with the installed material handling systems may cause excessive energy consumption and consequently CO_2 emission. In order to restrict the energy consumption and CO_2 emission to a reasonable level, a new implementation for material handling systems is required. In the European Union there are legalized regulations like Directive of the European Parliament and of the Council 2006/32/WE from 5th April, 2006 on energy end-use efficiency and energy services and repealing Council Directive 93/76/EEC; renewed Lisbon Strategy and National Cohesion Strategy (2007–2013). Also the Kyoto protocol somehow "commands" the signed countries to lower the CO_2 emissions for 20 % until the year 2020.

As mentioned previously, the throughput rate of a SBAWS is relatively high which may mean that there may be lots of movements and travels along the storage racks. Therefore, it is often likely to happen braking of lifting tables and shuttle carriers. One way to recover the energy is during braking of the lifting tables and the shuttle carriers. In order to calculate the energy regeneration, the knowledge in mechanic and kinematics is used.

When developing the model, the following assumptions and notations are considered:

- The warehouse is divided into picking aisles with storage racks on both sides; therefore there are double storage racks between the picking aisle and single storage rack along the warehouse walls. The input/output ($I/O_{aisle i}$) location of the warehouse is located at the lower left side of each elevator (see Figure 2).
- The storage rack has a rectangular shape and is divided by levels. At the beginning of the storage rack, there is a buffer area. In each level *i*, there is a single shuttle carrier – aisle captive shuttle carrier (see Figure 2).
- The elevator manipulates independently two lifting tables one of which is located at the left side and the other one is located at the right side of the elevator. Each lifting table can serve one container at a time.
- The elevator with the lifting table enables the operation of single and dual command cycles.
- The shuttle carrier enables the operation of the single and dual command cycles.
- Drive characteristics of the elevator (lifting tables) as well as the height of the storage rack are known priori.
- Drive characteristics of the shuttle carriers as well as the length of the storage racks are known priori.
- The height H_{SR} and length L_{SR} of the storage racks are large enough for the elevator and shuttle carriers to reach its maximum velocity v_{max} in the vertical and in the horizontal direction.
- Randomized policy is used, which means that any buffer position at level *i* is equally likely to be selected for the storage or retrieval assignment by the elevator.
- Randomized policy is used, which means that any rack opening in the storage compartment is equally likely to be selected for the storage or retrieval assignment by the shuttle carrier.

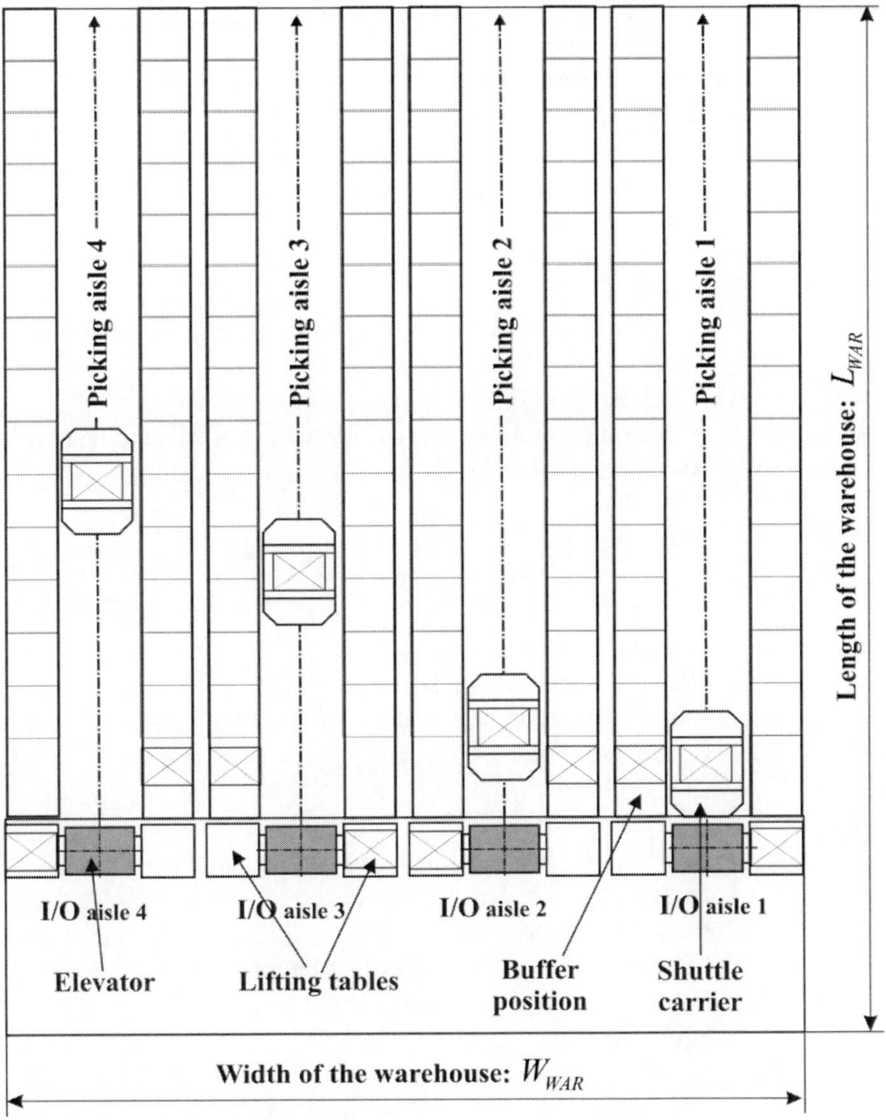

Figure 2. The layout of the shuttle based automated warehouse.

Also the following notation is introduced:

AS/RS	Automated storage and retrieval systems.
AVS/RS	Autonomous vehicle storage and retrieval system.
CO_2	Carbon dioxide.

FEM	European federation of materials handling.
F.F.	Football field.
I/O	Input and output location.
ROI	Return of investment.
SBAWS	Shuttle-based automated warehouse system.
S/R	Storage and retrieval.
a	Acceleration/deceleration.
a_y	Acceleration/deceleration of the lifting table in the vertical direction.
a_x	Acceleration/deceleration of the shuttle carrier in the horizontal direction.
a_y^+	Acceleration of the lifting table in the vertical direction.
a_x^+	Acceleration of the shuttle carrier in the horizontal direction.
g	Acceleration of gravity.
F_A	Air resistance force.
T(SC)	Average single command cycle time.
T(DC)	Average dual command cycle time.
F_B	Braking force.
M_B	Braking torque.
l	Braking distance of the shuttle carrier.
h	Braking height of the elevators lifting table.
C	Center of gravity.
E_{CO2}	CO_2 emissions.
a_y^-	Deceleration of the lifting table in the vertical direction.
a_x^-	Deceleration of the shuttle carrier in the horizontal direction.
s	Distance.
ER	Energy recovery.
ε	Efficiency of the warehouse (distribution center).
$E(t_1)$	Expected single command cycle time.
$E(t_2)$	Expected dual command cycle time.
k_{ir}	Factor which take into account the impact resistance of rotating masses with variable vehicle velocity.
ρ	Factor of emission which is obtained based on measurements.
G	Force of gravity ($G = m_g \cdot g$).
G_B	Force of gravity of the front wheel.
G_A	Force of gravity of the rear wheel.
F_R	Force of rolling friction.
h	Height of the storage rack level (storage compartment).

H_{SR}	Height of the storage rack.
h	Height of the container.
h_{COM}	Height of the storage compartment.
Q	Heat flow.
E_k	Kinetic energy.
L_{SR}	Length of the storage rack.
l	Length of the container.
l_{COM}	Length of the storage compartment.
L_{WAR}	Length of the warehouse.
H	Lifting height of the elevator.
m_g	Mass.
v_y	Maximum velocity of the hoisted carriage in the vertical direction.
v_x	Maximum velocity of the S/R machine in the horizontal direction.
v_{max}	Maximum velocity.
n	Number of containers in storage compartment.
m	Number of levels.
N_x	Number of storage compartments in the horizontal direction,
N_y	Number of storage compartments in the vertical direction,
n_{weeks}	Number of weeks.
n_{wd}	Number of working days in a week.
n_{DC}	Number of cycles.
T_{shift}	Number of working hours in a shift.
t_l	Pick up/deposit time.
E_p	Potential energy.
S_{Forest}	Required amount of forest to neutralize CO_2 emissions in 10 years.
$S_{Footbal}$	Required amount of football field to neutralize CO_2 emissions in 10 years.
k_r	Rolling resistance coefficient.
ΣF_x	Sum of all the forces in the horizontal direction.
ΣF_y	Sum of all the forces in the vertical direction.
ΣM_i	Sum of all moments.
Pf	Throughput performance.
λ(DC)	Throughput capacity of dual command cycle.
T	Time.
$T(s)$	Time in dependence of distance.
$t_{\mathrm{I/O},j}$	Travel time of the lifting table from I/O to j.

$t_{i,j}$	Travel time of the lifting table from *i* to *j*.
$t_{\text{I/O level } j,\text{ P1}}$	Travel time of the shuttle carrier from the I/O$_{\text{level}j}$ to point P_1.
$t_{\text{P1, P2}}$	Travel time of the shuttle carrier from the P_1 to point P_2.
$t_{\text{P1, I/O level } j}$	Travel time of the shuttle carrier from P_2 to the I/O$_{\text{level}j}$.
F_T	Traction force.
$v(t)$	Velocity in dependence of time.
vp_i	Velocity profile.
W	Work.
w	Width of the container.
w_{COM}	Width of the storage compartment.
W_{WAR}	Width of the warehouse.

4. Conceptual Model for Calculating the System Performance of the SBAWS

The proposed conceptual model for calculating the system performance of the SBAWS encompasses (*i*) travel time definition (*ii*) throughput performance of the elevator, (*iii*) throughput performance of the shuttle carrier and (*iv*) energy recovery model of the elevator and the shuttle carrier.

4.1. Travel Time Definition

When developing the conceptual model for the proposed SBAWS, the real operating characteristics of the elevators and the shuttle carriers are considered. The analytical travel time model based on the assumption of uniform velocity is too general. If we limit ourselves in using the uniform velocity only, the average travel time for a cycle becomes shorter and consequently the throughput capacities become larger, which is not in accordance with the practice. The above-mentioned statement becomes even more evident by short distances of the storage racks.

Generally, two types of velocity profiles can be distinguished depending on whether the obtained peak velocity $v(t_p)$ is less than $v_{\max}$(type I) or equal to $v_{\max}$ (type II) (Figure 3). It can be verified that time $T < 2v_{\max}/a$ for type I and $T > 2v_{\max}/a$ for type II.

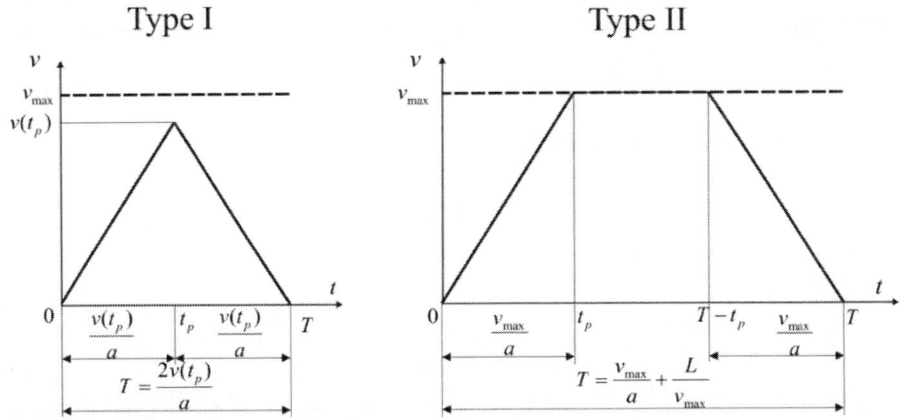

Figure 3. Velocity – time relationship.

4.1.1. Velocity Profile Type I

The velocity in dependence of time $v(t)$ is calculated by (1):

$$v(t)=\begin{cases} at, & t\in(0,t_p) \\ -a(t-T), & t\in(t_p,T) \end{cases} \tag{1}$$

The distance in dependence of time $s(T)$ is calculated by (2):

$$s(T)=\int_0^T v(t)\,dt \tag{2}$$

Considering the condition $t_p = T/2$, the distance in dependence of time $s(T)$ is calculated by (3):

$$s(T)=\frac{a\cdot T^2}{4} \tag{3}$$

Time in dependence of distance $T(s)$ is obtained by (4):

$$T(s) = 2 \cdot \sqrt{\frac{s}{a}} \tag{4}$$

4.1.2. Velocity Profile Type II

The velocity in dependence of time $v(t)$ is calculated by (5):

$$v(t) = \begin{cases} at, & t \in (0, t_p) \\ v_{\max}, & t \in (t_p, T - t_p) \\ -a(t - T), & t \in (T - t_p, T) \end{cases} \tag{5}$$

The distance in dependence of time $d(T)$ is calculated by (6):

$$d(T) = \int_0^T v(t)\,dt \tag{6}$$

Considering the condition$t_p = v/a$, the distance in dependence of time $s(T)$ is calculated by (7):

$$d(T) = v \cdot T - \frac{v^2}{a} \tag{7}$$

Time in dependence of distance $T(s)$ is calculated by (8):

$$T(s) = \frac{s}{v} + \frac{v}{a} \tag{8}$$

According to the (4) and (8), time in dependence of distance $T(s)$ is calculated by (9):

$$T(s)=\begin{cases} 2\cdot\sqrt{\dfrac{s}{a}}, & s\le\dfrac{v^2}{2a} \\ \dfrac{s}{v}+\dfrac{v}{a}, & s\ge\dfrac{v^2}{2a} \end{cases} \tag{9}$$

4.2. Throughput Performance of the Elevator

Recall that there are two lifting tables at the elevator. The elevator with lifting tables is feeding the storage racks by moving containers to buffer positions, which are located at the front of each level of the storage rack. At each level, there is a shuttle carrier that transports containers between the buffer position and storage locations and vice versa. The buffer position has only one transferring point, where storage request are delivered and retrieval request are picked up by the lifting table.

The elevator can work on a single command mode, which means that only one container is handled in a cycle. More advanced is double command mode, by which two containers are handled in a cycle.

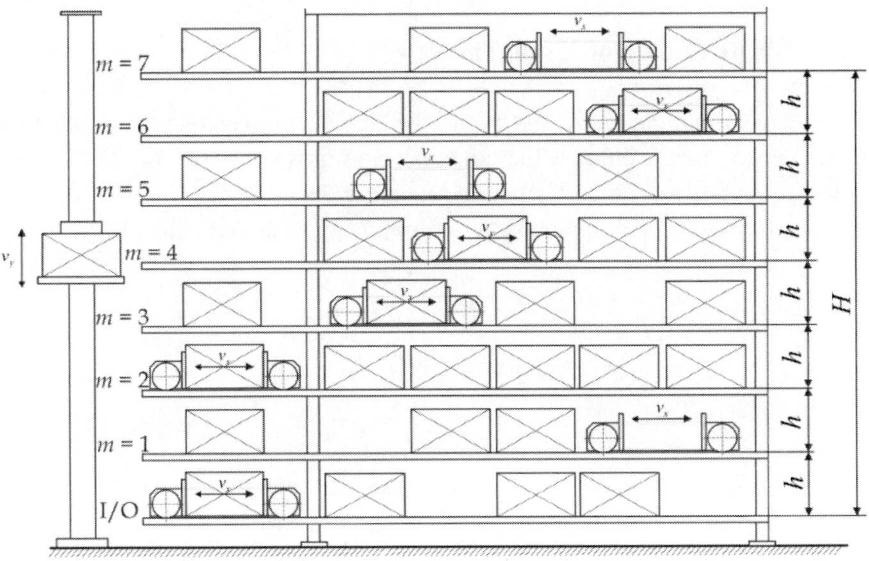

Figure 4. Shuttle based automated warehouse system with *m* levels.

Figure 4 shows an example for a SBAWS with m levels operated by the elevators. Elevators can serve m levels each one with an average height of h (see Figure 4).

By assuming the distances between levels to be large enough for the lifting table to achieve its maximum velocity v_{max} and a continuous sequence of loaded and unloaded moves between destination points, average cycle time using the probability theory, can be calculated (Oser and Reisinger, 2008).

Travel time from I/O location to the buffer position at level j is calculated by (10):

$$t_{\mathrm{I/O},j} = \frac{v_y}{a_y} + \frac{H}{m}\frac{(m+1)}{2v_y} \quad \text{travel time from I/O to } j \tag{10}$$

Travel time from the buffer position at level i to the buffer position at level j, is calculated by (11):

$$t_{i,j} = \frac{v_y}{a_y} + \frac{H}{m}\frac{(m+1)}{3v_y} \quad \text{travel time from } i \text{ to } j \tag{11}$$

Expected Single Command Cycle Time

In the case of single command cycle, the elevator moves the lifting table with the container to level j, unloads the container and moves back to the I/O location. The same sequence can be performed in reverse order.

The expected single command cycle time $E(t_1)$ is calculated by (12):

$$\begin{aligned} E(t_1) &= 2 \cdot t_l + 2 \cdot t_{\mathrm{I/O},j} \\ E(t_1) &= 2 \cdot t_l + 2\frac{v_y}{a_y} + \frac{h}{v_y}(m+1) \end{aligned} \tag{12}$$

Expected Dual Command Cycle Time

In the case of dual command cycle, the elevator moves the lifting table with the container to the level i, unloads the container and moves further to the level j, where the container is retrieved. After loading the container at level j, the elevator moves the lifting table back to the I/O location.

The expected dual command cycle time $E(t_2)$ is calculated by (13):

$$E(t_2) = 4 \cdot t_l + 2 \cdot t_{\mathrm{I/O},j} + 2 \cdot t_{i,j}$$
$$E(t_2) = 4 \cdot t_l + 2\frac{v_y}{a_y} + \frac{h}{v_y}(m+1) + 2\frac{v_y}{a_y} + \frac{2}{3}\frac{h}{v_y}(m+1) \tag{13}$$

By considering (12) and (13) the throughput performance per hour is obtained by (14):

$$Pf = \frac{3600}{E(t_i)} \cdot k \quad (k = 1 \text{ for } i = 1 \text{ or } 2 \text{ for } i = 2) \tag{14}$$

4.3. Throughput Performance of the Shuttle Carrier

A shuttle carrier can receive one container at a time and can operate on a single or double command modes (FEM Section IX, 2001).

Travel time of the shuttle carrier from the $\mathrm{I/O}_{\mathrm{level}j}$ to point P_1 is calculated by (15) (see Figure 5):

$$t_{\mathrm{I/O}_{\mathrm{level\,j}}, P_1} = \frac{v_x}{a_x} + \frac{x_1 - x_0}{v_x}$$
$$t_{\mathrm{I/O}_{\mathrm{level\,j}}, P_1} = \frac{v_x}{a_x} + \frac{1L}{5v_x} \tag{15}$$

Travel time of the shuttle carrier from the point P_1 to the point P_2 is calculated by (16) (see Figure 5):

$$t_{P_1, P_2} = \frac{v_x}{a_x} + \frac{x_2 - x_1}{v_x}$$
$$t_{P_1, P_2} = \frac{v_x}{a_x} + \frac{7L}{15v_x} \tag{16}$$

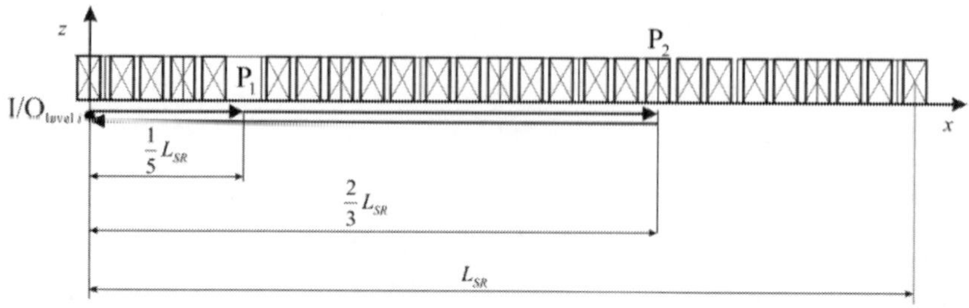

Figure 5. Modified FEM 9.851 for the shuttle carrier movement in the horizontal direction.

Travel time of the shuttle carrier from the point P_2 to I/O_{levelj} is calculated by (17) (see Figure 5):

$$t_{P_2,\, I/O_{level\ j}} = \frac{v_x}{a_x} + \frac{x_2 - x_0}{v_x}$$

$$t_{P_2,\, I/O_{level\ j}} = \frac{v_x}{a_x} + \frac{2L}{3v_x} \tag{17}$$

Average Single Command Cycle Time

The operation of the single command mode encompasses either the storage or the retrieval sequence. After completing a given storage or retrieval request at point $P_i(x_i)$, the shuttle carrier returns directly to the I/O_{levelj}. According to the Figure 5, a single storage request is performed at point P_1 and a single retrieval request is performed at point P_2. In the case of the single command, the shuttle carrier operates only one container.

Average single command cycle time $T(SC)$ is calculated by (18):

$$T(\mathrm{SC}) = 2 \cdot t_l + \frac{\left(t_{I/O_{level\ j},P_1} + t_{P_2,\, I/O_{level\ j}}\right)}{2} \tag{18}$$

Average Dual Command Cycle Time

The operation of the double command mode considers the storage and the retrieval sequences. In the dual command cycle, the shuttle carrier visits up to two storage locations between successive returns to the I/O_{levelj}. After completing a given storage request at point $P_1(x_1)$, the shuttle carrier moves directly to another storage location for the next retrieval request at point $P_2(x_2)$ without returning to the I/O_{levelj}. In the double command mode, the shuttle carrier carries two containers.

Average dual command cycle time T(DC) is calculated by (19):

$$T(DC) = 4 \cdot t_l + t_{\mathrm{I/O_{level\,j}},\mathrm{P_1}} + t_{\mathrm{P_1},\mathrm{P_2}} + t_{\mathrm{P_2},\mathrm{I/O_{level\,j}}} \tag{19}$$

According to (18) and (19) the throughput is calculated by (20):

$$Pf = \frac{3600}{T(i)} \cdot k \qquad \left(k = 1 \text{ for } i = \mathrm{SC} \text{ or } 2 \text{ for } i = \mathrm{DC}\right) \tag{20}$$

4.3. Energy Regeneration

Warehouses and distribution centers are relatively large energy consumers. According to the recent trends, warehouses along with the material handling equipment are following the goal of the energy efficiency. One way which is very attractive in practice is the energy regeneration during braking of the elevators (moving to the I/O location) and braking of the shuttle carriers.

In the following sections, calculations for the energy regeneration of the elevators and the shuttle carriers are presented. Also, the amount of the CO_2 emission that could be saved using the concept of energy recovery is presented.

4.3.1. Energy Regeneration During Braking of the Shuttle Carrier

If the body (in our case shuttle carrier with the container – see Figure 6) is in balance, then (21) is valid:

$$\begin{aligned} &\Sigma F = 0,\ \Sigma M_i = 0 \\ &\Sigma F_x = 0,\ \Sigma F_y = 0,\ \Sigma M_i = 0 \end{aligned} \tag{21}$$

When the sum of all the forces in the horizontal direction ΣF_x equals zero, then (22) is valid:

$$\begin{aligned}&\Sigma F_x = 0\\&F_R + F_A - F_{iT} + F_B = 0\\&F_B = F_{iT} - F_R - F_A\end{aligned} \tag{22}$$

When the sum of all the forces in the vertical direction ΣF_y equals zero, then (23) is valid:

$$\begin{aligned}&\Sigma F_y = 0\\&G_A + G_B - G = 0\\&G = G_A + G_B\end{aligned} \tag{23}$$

The aerodynamic drag F_A is assumed to have no influence on the traction force F_T. Since the shuttle carrier travels on leveled surfaces only, the climbing resistance is equal to zero and is therefore not considered in the model.

Rolling resistance F_R equals (24):

$$\begin{aligned}&F_R = (G_A + G_B)\cdot k_r\\&F_R = G\cdot k_r\end{aligned} \tag{24}$$

Inertial resistance F_{iT} equals (25):

$$\begin{aligned}&F_{iT} = m_g \cdot a_x^- \cdot k_{ir}\\&F_{iT} = \frac{G}{g}\cdot a_x^- \cdot k_{ir}\end{aligned} \tag{25}$$

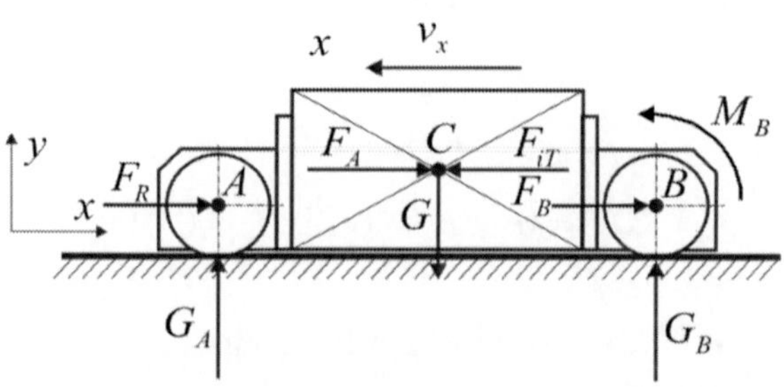

Figure 6. The activity of all forces and the traction moment in case of braking of the shuttle carrier.

In the case of braking of the shuttle carriers, the braking force F_B on the rear driving wheel equals (26):

$$F_B = \frac{G}{g} \cdot a_x^- \cdot k_{ir} - G \cdot k_r \tag{26}$$

It can be noticed that the braking force F_B on the rear driving wheel is overcoming the inertial resistance F_{iT}, which is reduced for the rolling resistance F_R. The size of the braking force F_B depends on: (*i*) mass of the shuttle carrier with the container, (*ii*) size of deceleration a_x^-, (*iii*) factor k_{ir} that takes into account the resistance of rotating masses with variable vehicle speed and (*iv*) rolling resistance coefficient k_r.

When shuttle carrier starts braking, it has the kinetic energy E_k. It is assumed that all the kinetic energy E_k is transformed to work W without any losses for the heat flow Q into the system, which equals (27):

$$W = W_B = F_B \cdot l \; [\mathrm{J}] \tag{27}$$

The symbol l represents the braking distance of the shuttle carrier and equals (28):

$$l = \frac{v_x^2}{2 \cdot a_x^-} \tag{28}$$

4.3.2. Energy Regeneration During Braking of the Elevators Lifting Table

If the body (in our case the lifting table with the container – see Figure 7) is in balance, then (29) is valid:

$$\begin{aligned} &\Sigma F = 0,\ \Sigma M_i = 0 \\ &\Sigma F_x = 0,\ \Sigma F_y = 0,\ \Sigma M_i = 0 \end{aligned} \tag{29}$$

When the sum of all the forces in the vertical direction ΣF_y equals zero, then (30) is valid:

$$\begin{aligned} &\Sigma F_y = 0 \\ &F_B - G - F_{iT} + F_A = 0 \\ &F_B = G + F_{iT} - F_A \end{aligned} \tag{30}$$

The aerodynamic drag F_A and the rolling resistance F_R are assumed to have no influence on the traction force F_T. Inertial resistance F_{iT} equals (31):

$$\begin{aligned} &F_{iT} = m_g \cdot a_y^- \cdot k_{ir} \\ &F_{iT} = \frac{G}{g} \cdot a_y^- \cdot k_{ir} \end{aligned} \tag{31}$$

When the lifting table decelerates ($a_y^+ \neq 0$), the braking force F_B equals (32):

$$F_B = G + \frac{G}{g} \cdot a_y^- \cdot k_{ir} \tag{32}$$

It can be noticed that the braking force F_B is overcoming the force of gravity G and the inertial resistance F_{iT}. The size of the braking force F_B depends on the: (*i*)mass of the lifting table along with the container, (*ii*) size of acceleration a_x^+ and (*iii*) factor k_{ir} that takes into account the resistance of rotating masses with variable vehicle speed.

When lifting table starts braking, it has the potential E_p and the kinetic energy E_k. It is assumed that all the potential energy E_p and the kinetic energy E_k are transformed to work W without any losses for the heat flow Q into the system, which equals (33):

$$W = W_B = F_B \cdot h \ [\mathrm{J}] \tag{33}$$

The symbol h represents the braking height of the lifting table and equals (34):

$$h = \frac{v_y^2}{2 \cdot a_y^-} \tag{34}$$

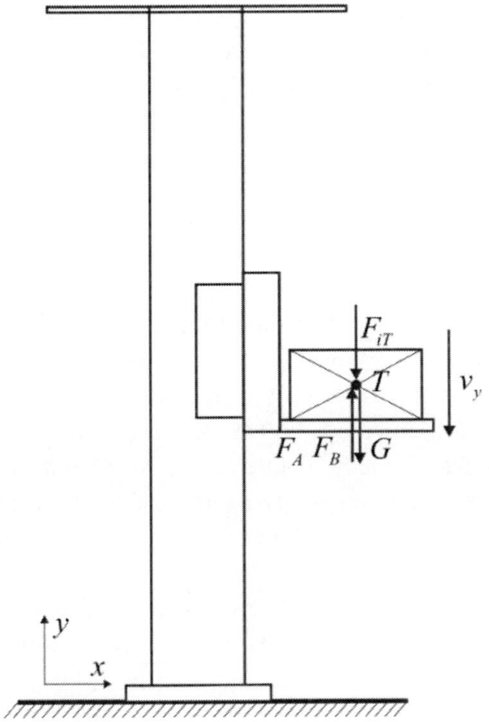

Figure 7. The activity of all forces in case of braking of the elevators lifting table.

4.3.3. Calculation of the Energy Regeneration

Energy regeneration in the case of braking of the shuttle carrier depends on the kinetic energy E_k which is transformed to work W without any losses for the heat flow Q into the system. In order to define the number of brakes of the shuttle carrier in the prescribe level of the storage rack the number of single or dual command cycles is required to be calculated. In this study, dual command cycle is considered.

According to the dual command cycle, three independent routes for the shuttle carrier should be considered (see Figure 5):

- traveling from the I/O_{levelj} to point P_1,
- traveling from point P_1 to point P_2,
- traveling from point P_2 to the I/O_{levelj}.

It is assumed, that the energy regeneration during braking of the shuttle carrier is accrued in all three independent routes of the shuttle carrier ($z = 3$). According to this assumption, the work W equals (35):

$$W = W_B = F_B \cdot l \cdot z \cdot 2.78 \cdot 10^{-7} \text{ (kWh)} \tag{35}$$

In order to express the work W in ***kilowatt hour*** (kWh) instead of in joule (J), the following relationship 1 J = 2.78 · 10^{-7} kWh has been considered.

Energy recovery in case of braking of the elevator lifting table depends on the potential energy E_p and the kinetic energy E_k which are transformed to work W without any loss for the heat flow Q in the system. As mentioned previously, in order to calculate the number of brakes of the elevator lifting table, number of single or dual command cycle is required to be calculated. Because dual command cycle is considered in the study, two independent movements of the elevator lifting table are required to be calculated:

- movement of the elevator lifting table from level j (storage request) to level i (retrieval request),
- movement of the elevator lifting table from level i (retrieval request) to the I/O location.

It is assumed that the energy regeneration during braking of the elevator lifting table is accrued in only one case which corresponds to movement of the elevator lifting table from level i (retrieval request) to the I/O location($z = 1$). According to this assumption, the work W equals (36):

$$W = W_B = F_B \cdot h \cdot z \cdot 2.78 \cdot 10^{-7} \text{ (kWh)} \tag{36}$$

In order to express the work - W - in ***kilowatt hour*** (kWh) instead of in joule (J), the following relationship 1 J = 2.78 · 10^{-7} kWh is considered.

Energy regeneration ER counted on a yearly basis, depends on the energy regeneration of all shuttle vehicles and elevator lifting tables during operation of dual command cycle, the number of dual command cycles n_{DC} of the elevator and the shuttle carrier, the number of working hours per shift T_{shift}, the number of working days per week n_{wd}, the number of weeks n_{weeks}, the efficiency of the shuttle based automated warehouse ε and it is calculated by (37):

$$ER = W_B \cdot n_{DC} \cdot T_{shift} \cdot n_{wd} \cdot n_{weeks} \cdot \varepsilon \left[\frac{\text{kWh}}{\text{year}} \right] \tag{37}$$

By the energy regeneration contribution, the amount of CO_2 emission saved from going into the atmosphere can be calculated by (38):

$$E_{CO2} = W \cdot \rho \left[\frac{\text{kgCO}_2}{\text{year}} \right] \tag{38}$$

The symbol ρ stands for the factor which is obtained based on measurements. In our case we refer the factor ρ of the Umweltbundesamt from Germany.

In nature, 100 m^2 of the forest consume approximately 1 tone of CO_2 within 10 years for the operation of photosynthesis from the atmosphere. This relationship can be applied when we would like to express CO_2 emission in the atmosphere with the required amount of forest that could be saved using the concept of energy recovery.

If the amount of the CO_2 emissions saved from going into the atmosphere is measured in tCO_2/year, then (39) is valid.

$$S_{Forest} = E_{CO_2} \cdot 100\, m^2 \left[\frac{\text{m}^2 \text{of Forest}}{10 \text{ year}} \right] \tag{39}$$

If the amount of the CO_2 emissions saved from going into the atmosphere is measured in $kgCO_2$/year then (40) is valid.

$$S_{Forest} = E_{CO_2} \cdot 0.1\, m^2 \left[\frac{\text{m}^2 \text{of Forest}}{10 \text{ year}} \right] \tag{40}$$

In order to present the S_{Forest} in hectares, the result of (39) must be divided by 10.000. In order to present S_{Forest} in pieces of football fields, the result of (39) must be divided by 7.140.

$$S_{Footbal} = \frac{S_{Forest}}{7140} \left[\frac{\text{piece of F.F.}}{10 \text{ year}} \right] \tag{41}$$

5. SBAWS under Study

In this chapter, generic approach for throughput performance and energy efficiency calculations for SBAWS is presented. Generally the efficiency of a SBAWS depends on the efficiency (velocity profile vp_i) of the elevators and the efficiency of the shuttle carriers, design of the storage racks (L_{SR} and H_{SR}) and the control policy. A case study will be provided in this section.

In the case study, stock keeping unit represents a plastic container filled with items with the dimensions: length $l = 0.6$ m, width $w = 0.4$ m and height $h = 0.24$ m. With regard to the plastic container, the storage area has the following dimensions: length $l_{COM} = 0.6$ m, width $w_{COM} = n$ x 0.5 m ($n = 5$) and height $h_{COM} = 0.3$ m. Dimensions of the storage rack (L_{SR} and H_{SR}) depends on the number of storage locations in the horizontal and vertical directions, N_x and N_y, respectively.

Eleven different types of SBAWs (see Table 1) are considered.

Table 1. SBAWSs under study

i	N_x	m $(N_y - 1)$	L_{SR} (m)	H_{SR} (m)
1	5	10	12.5	3.3
2	5	11	12.5	3.6
3	5	12	12.5	3.9
4	5	13	12.5	4.2
5	5	14	12.5	4.5
6	5	15	12.5	4.8
7	5	16	12.5	5.1
8	5	17	12.5	5.4
9	5	18	12.5	5.7
10	5	19	12.5	6.0
11	5	20	12.5	6.3

Based on the efficiency of the elevator with the lifting table and the shuttle carriers the combination of eleven different velocity scenarios are considered (see Table 2).

The total mass of the lifting table, shuttle carrier and the container are presented in Table 3.

Table 2. Velocity scenarios of the shuttle carriers and the elevators lifting table

Velocity profile	Shuttle carrier travelling in the horizontal direction			Lifting table movement in the vertical direction		
	v_x (m/s)	a_x^+(m/s^2)	a_x^-(m/s^2)	v_y(m/s)	a_y^+(m/s^2)	a_y^-(m/s^2)
vp_1	1.5	0.75	0.75	1.5	1.5	1.5
vp_2	1.5	1.0	1.0	1.5	1.5	1.5
vp_3	1.5	1.5	1.5	1.5	1.5	1.5
vp_4	2.0	1.0	1.0	1.5	1.5	1.5
vp_5	2.0	1.5	1.5	1.5	1.5	1.5
vp_6	2.0	2.0	2.0	1.5	1.5	1.5
vp_7	3.0	1.5	1.5	1.5	1.5	1.5
vp_8	3.0	2.0	2.0	1.5	1.5	1.5
vp_9	3.0	2.25	2.25	1.5	1.5	1.5
vp_{10}	3.0	3.0	3.0	1.5	1.5	1.5
vp_{11}	4.0	3.0	3.0	1.5	1.5	1.5

Note: Velocity scenarios from vp_1 to vp_{11} are selected according to the references of material handling equipment producers and practical experiences of the author.

Table 3.Total mass of the lifting table, shuttle carrier and the container

mass of the full container (kg)	mass of the lifting table (kg)	mass of the shuttle carrier (kg)	total mass of the shuttle carrier (kg)	total mass of the lifting table (kg)
20	60	40	60	80

Note: The proposed masses are approximately specified according to the knowledge in material handling and might differ from the actual systems installed in practice.

Like different types of SBAWSs and velocity scenarios, the control policy has also a significant share on the travel time. In the proposed case study, the randomized storage and retrieval assignment policy with dual command cycle is applied. Along with the calculation of the mean travel time, additional time for manipulation of the container is used (t_l = 1.5 sec.).

The energy regeneration *ER* is calculated based on these factors: (*i*) work *W* measured in kWh as a consequence of braking of the elevator lifting table and the shuttle carrier, (*ii*) the number of cycles n_{DC}, (*iii*) the number of working hours per shift T_{shift} = 16 hours, (*iv*) the number of working days per week n_{wd} = 5, (*v*) the number of weeks n_{weeks} = 50, (*vi*) the efficiency of the warehouse (distribution centre) ε = 0.95. The CO_2 emission saved from going into the atmosphere is calculated by considering the emission factor as 0.59. The emission factor is obtained from the measurements of the Umweltbundesampt in Germany. The CO_2 emission saved from going into the atmosphere is calculated in terms of amount of forest measured in m^2 per 10 years or number of football fields per 10 years.

6. Analyses and Evaluation of Results

In the following sub-sections, results for: (*i*) average travel time T(DC) and throughput capacities λ(DC)for the elevator and the shuttle carrier, (*ii*) energy regeneration ER, (*iii*) CO_2 emissionE_{CO2} saved from going into the atmosphere and (*iv*) required amount of forest S_{Forest} saved due to energy regeneration will be presented and discussed. Results of the above mentioned factors are calculated by the proposed conceptual model of the SBAWS (see Section 4).

Table 4. Dual command cycle time and throughput capacity of the elevator

Velocity profile vp_i	No. of levels m	Dual command cycle time T(DC)	Throughput capacity λ(DC)
1	10	13.67*	527**
2	11	14.00	514
3	12	14.33	502
4	13	14.67	491
5	14	15.00	480
6	15	15.33	470
7	16	15.67	460
8	17	16.00	450
9	18	16.33	441
10	19	16.67	432
11	20	17.00	424

Note: *The dual command cycle time associated with load handling. **Throughput capacity in case of dual command cycle.

6.1. Throughput Performance of the Elevator

Dual command cycle time T(DC) and throughput capacities λ(DC) which are based on the data presented in Table 4 are given on the basis of the performed analyses. Due to the restriction on the chapter length, eleven different levels for the storage rack are presented as in Table 4. The relationship between the T(DC) and the λ(DC) is presented in Figure 8. The velocity of the elevators is set to v_y = 1.5 m/s and a = 1.5 m/s^2 and stands for all presented velocity scenarios.

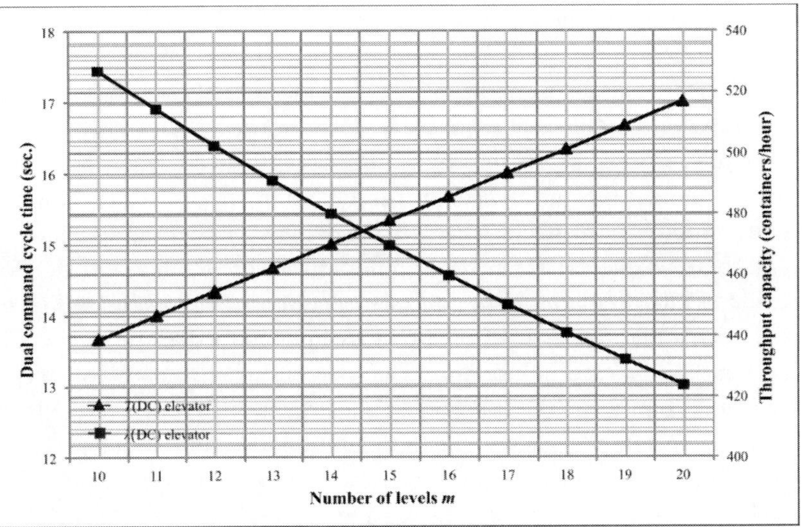

Figure 8. Distribution of the dual command cycle time and throughput capacity of the elevator.

According to the distribution of the T(DC), the number of levels m has significant impact on the cycle time (see Figure 8). The lowest cycle time belong to the storage rack with m = 10 levels, while the highest cycle time belong to the storage rack with m = 20 levels. The relationship between the T(DC) and m is somehow linear.

Because the throughput capacity λ(DC) is inversly dependent on the travel time, the highest throughput capacity λ(DC) can be obtained by the number of levels m = 10. On the contrary the lowest throughput capacity λ(DC) can be obtained by the number of levels m = 20.

It is generally known that before selecting a type of elevator the number of levels m provides significant information for calculating the elevator performance. Based on the performed analysis, the assumption from the industry, that the shuttle based automated warehouses could reach up to 1000 container per hour, is proven. For example, if we have a storage rack with m = 15 levels, then the elevator with two independent lifting tables could operate with the throughput capacity of λ(DC) = 940 containers per hour (470 containers per hour multiplied by 2 – see Table 4).

6.2. Throughput Performance of the Shuttle Carrier

Dual command cycle time T(DC) and throughput capacities λ(DC) which are based on the data presented in Table 5 are given on the basis of the performed analyses. Eleven different velocity scenarios vp_i for the shuttle carriers with the fixed number of levels m = 15 are presented as in Table 5. The relationship between the T(DC) and the λ(DC) is presented in Figure 9.

According to the distribution of the T(DC), velocity vp_i has a significant impact on the cycle time (see Figure 9). In T(DC) distribution significant changes of T(DC) are observed for the first three points (vp_1, vp_2 and vp_3). This relationship shows the influence of acceleration a_x^+, deceleration a_x^- and horizontal velocity v_x in accordance to the length of the storage rack L_{SR}. A slightly decrease of the T(DC) arises at point vp_4, where the horizontal velocity v_x is increased, while acceleration a_x^+ and deceleration a_x^- are decreased (see Table 2).This confirms that for such storage rack designs the values of acceleration a_x^+ and deceleration a_x^- are important. The decreasing tendency of the T(DC) continues until point vp_6. T(DC) becomes higher at point vp_7 where the horizontal velocity v_x is increased, meanwhile acceleration a_x^+ and deceleration a_x^- are decreased (See Table 2). Increased horizontal velocity v_x, acceleration a_x^+ and deceleration a_x^- cause once again a decreasing tendency of T(DC) for other points vp_i. Generally, the best results are achieved by the shuttle carriers having fast drivers in the horizontal travelling direction. The lowest T(DC) belongs to the velocity profiles vp_{10} and vp_{11} which currently represent the state of the art technology in warehousing.

Because of the throughput capacity λ(DC) is inversly dependent on the travel time T(DC), the highest λ(DC) can be obtained by the velocity vp_{11}. On the contrary the lowest λ(DC) can be noticed by the velocity vp_1.

Generally, when selecting the type of a shuttle carrier, velocity vp_i provides significant information in calculating the shuttle carrier's efficiency. According to the relatively high throughput capacity λ(DC) of a shuttle carrier, the assumption that in each level there is a single shuttle carrier could be relased. For example, if we have a storage rack with m = 15 levels and the elevator with two individual lifting tables, the throughput of the system could be near to 1000 containers per hour. Since we have a system with a single shuttle carrier in each level, the throughput capacity 1000 containers per hour is divided by 15 which gives us 67 containers per hour (n_{DC} = 34 cycles). The throughput capacity of the "slowest" shuttle carier is 247 containers per hour (see Table 5) which proves the assumption that the shuttle carrier could be operated in more levels within an aisle.

Table 5. Dual command cycle time and throughput capacity of the shuttle carrier

Velocity profile vp_i	No. of levels m	Dual command cycle time T(DC)	Throughput capacity λ(DC)
1	15	29.11*	247**
2	15	27.61	261
3	15	26.11	276
4	15	26.33	273
5	15	24.33	296
6	15	23.33	309
7	15	23.56	306
8	15	22.06	326
9	15	21.56	334
10	15	20.56	350
11	15	20.17	357

Note: *The dual command cycle time associated with load handling. **Throughput capacity in case of dual command cycle.

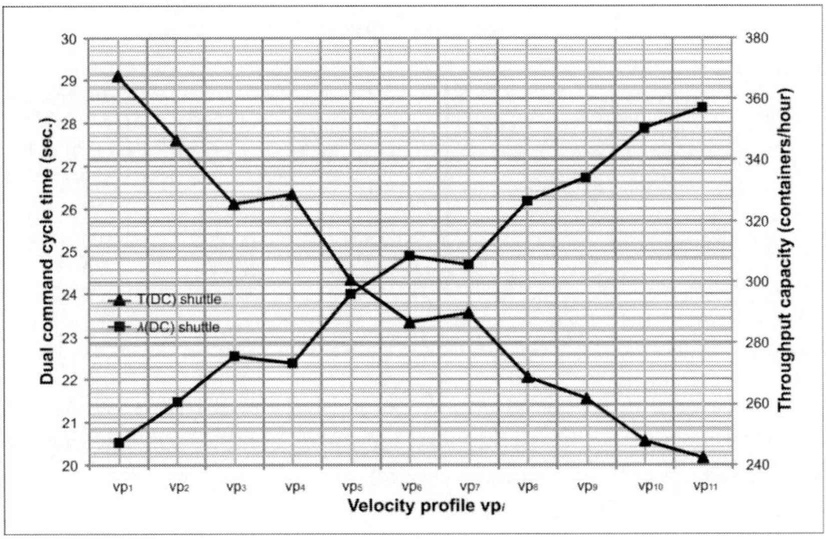

Figure 9. Distribution of the dual command cycle time and throughput capacity of the shuttle carrier.

Based on the performed analysis for the elevator and the shuttle carrier, it can be concluded that the throughput performance of the SBAWS is determined mostly by the elevator. Therefore, the elevator should receive the most of the attention when designing SBAWSs.

6.3. Energy Regeneration

The energy regeneration *ER*, CO_2 emission E_{CO2} saved from going into the atmosphere and the required amount of forest to neutralize CO_2 emissions in 10 years are presented in Table 6. These values are given on the basis of the performed analyses. Because of the expressive influence of the velocity profile vp_i on the above mentioned factors (*ER*, E_{CO2} and S_{Forest}), the analysis has been performed for the storage rack with m = 15 levels and selected 5 different velocity scenarios (vp_3, vp_5, vp_8, vp_{10} and vp_{11}). The relationship between the E_R and the R_{CO2}, is presented in Figure 10.

According to the distribution of the energy regeneration *ER* for the selected SBAWS, the velocity profile has a significant impact on the energy regeneration (see Figure 10).

Table 6. Energy regeneration, CO_2 emission and the required amount of forest

vp_i	E_R (kWh/ year)	E_{CO2} (kg CO^2/ year)	S_{Forest} (m^2 Forest/ 10 year)	S_{Forest} (ha Forest/ 10 year)	$S_{Football}$ (piece of F.F./ 10 year)
vp_3	492.66	290.67	29.07	0.0029	0.0041
vp_5	586.82	346.22	34.62	0.0035	0.0048
vp_8	856.57	505.38	50.54	0.0051	0.0071
vp_{10}	857.27	505.79	50.58	0.0051	0.0071
vp_{11}	1235.03	728.67	72.87	0.0073	0.0102

The lowest energy regeneration *ER* belongs to the SBAWS with relatively low velocity scenario, while the highest energy regeneration *ER* belongs to the SBAWS with extremly high velocity scenario.

According to the distribution of the energy regeneration *ER* for the SBAWS, the increasing tendency is observed for the first three points (vp_3, vp_5 and vp_8). This relationship shows a significant impact of the shuttle carrier velocity v_x, since the elevators lifting table velocity v_y remains the same for all the three points. The reader will notice that the energy regeneration *ER* is the same for points 3 and 4 (vp_8 and vp_{10}). This relationship can be explained with the fact, that velocities v_x and v_y are equal for velocity profiles vp_8 and vp_{10} (see Table 2). An increase of the energy regeneration *ER* arises at point 5 (vp_{11}), where the horizontal velocity is increased to $v_x = 4$ m/s.

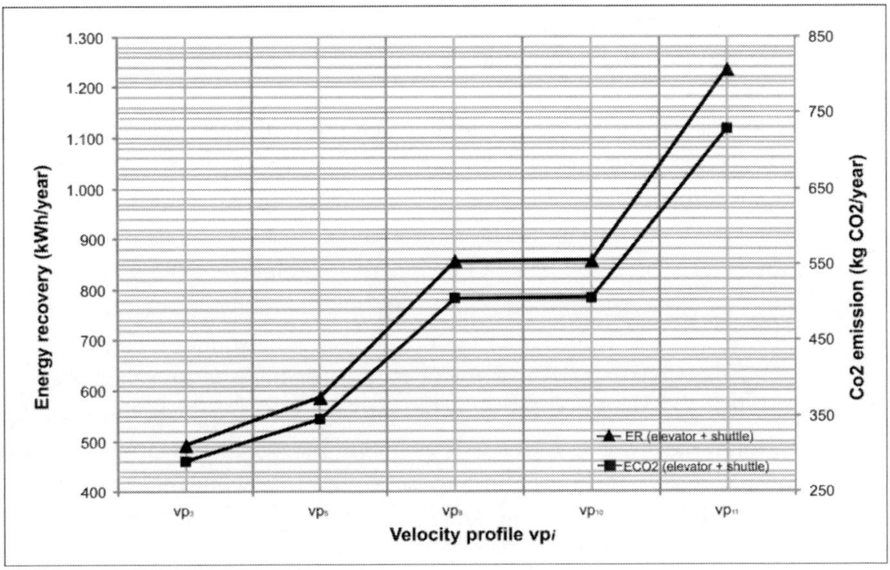

Figure 10. Distribution of the energy regeneration and CO_2 emission of the elevator lifting tables and shuttle vehicles.

Generally, the highest energy regeneration *ER* is achieved with the elevator and the shuttle carrier having the fastest drives. It can be concluded that the significant difference of energy regeneration *ER* in the present case originates as a result of changing the horizontal velocity v_x. The highest energy regeneration *ER* belongs to the velocity profile vp_{11}, which currently represents the state of the art technology in warehousing.

Since the CO_2 emission E_{CO2} saved from going into the atmosphere originates from the energy regeneration - *ER* - the relatively same dependance of E_{CO2} can be noticed in Figure 10. It can be seen that theoretically up to one tone of CO_2 could be saved from going into the atmosphere, if the energy regeneration is considered. Since there are lots of warehouses and distribution centers in the industry, this number could be relatively high. Consequently, if we consider environmentally friendly approaches and using the energy regeneration technology, a huge number of square meters or more appropriate square kilometers of forest can be saved.

Conclusion

In this chapter, a model for calculating the system performance of a SBAWS is presented. The proposed model consists of calculations for the throughput capacity of the elevator and the shuttle carriers and the calculations for the energy regeneration of the system.

SBAWs are usually used by companies dealing with large number of small sized orders requiring to be processed in a relatively short time period. In practice, there is an approximation that this system could reach up to 1000 containers per hour. This study fills the gap in the literature where there is no conceptual model calculating the system performance and the energy efficiency of the system.

The proposed conceptual model presents an analytical travel time model calculating the system performance of the elevator and the shuttle carrier.

Along with the system performance, a model for the energy regeneration is also proposed. It is believed that for an efficient SBAWS design, both throughput performance and energy regeneration aspect should be considered simultaneously.

Based on the results presented in Section 6, the following main conclusions can be summarized:

i. According to the results presented in Tables 4-5 and Figures 8-9, the throughput performance is influenced by the number of levels m, the velocity of the elevators lifting table (v_y, a) and the velocity of the shuttle carriers (v_x, a). The system performance mainly depends of the elevators feeding the storage racks. By this study, the assumption from the industry - the SBAWS could reach up to 1000 container per hour - is also proven. Of course, the above mentioned throughput performance is possibile, if the elevator can operate with two individual lifting tables. Since a shuttle carrier can handle more work than an elevator the assumption of aisle captive shuttles can be relased. In more advanced systems, a shuttle carrier can travel in more levels than a single one where it is transferred by a special lifting table (ussually attached an additional elevator at the back of the storage rack) to other levels. Since the throughput performance of a SBAWS is determined by the elevator, the elevator should receive the most of the attention when this system.

ii. According to the results presented in Table 6 and Figure 10, the energy regeneration is influenced by number of levels m and velocity of the elevators lifting table and the shuttle carriers (v_x, v_y and a). Generally, the highest energy regeneration ER is achieved with the elevator and the shuttle carriers with fast drives. Based on the application of the energy regeneration during braking of the elevators lifting tables and shuttle carriers, relatively high amount of CO_2 could be saved from going into the atmosphere. Consequently, a large number of square meters or more square kilometers of forest could be saved, if we were considering energy regeneration in those warehouses. It is believed that environment aspect will indubitably bring changes into planning of warehouses and will mean great challenge for those who are engaged in the planning process.

Acknowledgment

The author of this chapter contribution would like to express special thanks to Dr. Banu Yetkin Ekren and D.Sc. Zaki Sari as well as to all collaborators at the University of Maribor, Faculty of Mechanical Engineering who have in any way contributed to this chapter.

References

Carlo, H.J. & Vis, I.F.A. (2012).Sequencing dynamic storage systems with multiple lifts and shuttles. *Int. J. Production Economics*, (140), 844–853.

Ekren, B.Y. & Heragu, S.S. (2012). Performance comparison of two material handling systems: AVS/RS and AS/RS. *International Journal of Production Research*, (50), 4061-4074.

Ekren, B.Y., Heragu, S.S., Krishnamurthy, A. & Malmborg, C.J. (2012). Simulation based experimental design to identify factors affecting performance of AVS/RS. *Computers & Industrial Engineering*, (58), 175–185.

FEM 9.851. Performance data of storage and retrieval machines: Cycle times. 2011.

GEBHARDT. http://www.gebhardt-foerdertechnik.de/. Accessed 2th August 2013.

Kartnig, G., Grösel, B. & Zrnić, N. (2012).Past, state of the art and future of intralogistics. *Proceedings of the XX International Conference MHCL*, 12, 401–408.

Malmborg, C.J. (2002). Conceptualizing tools for autonomous vehicle storage and retrieval systems. *International Journal of Production Research*, 40 (8), 1807–1822.

Oser, J. & Reisinger, K. (2008). Cycle time and drive optimization of a high performance transfer car. Progress in Material Handling Research. *The Material Handling Industry of America*, (446 – 457).

Roodbergen, K.J. & Vis, F.A. (2009). A survey of the literature on automated storage and retrieval systems. *European Journal of Operational Research*, 194, 343–362.

TGW. http://www.tgw-group.com/. Accessed 2th August 2013.

UMWELTBUNDESAMPT. http://www.umweltbundesampt.de/. Accessed 2th August 2013.

In: Automation Systems of the 21st Century
Editors: D. Arent and M. Freebush
ISBN: 978-1-62948-262-0

Chapter 3

FLOW METHODOLOGIES FROM THE POINT OF VIEW OF GREEN ANALYTICAL CHEMISTRY

Antonio Ruiz-Medina* and Eulogio J. Llorent-Martínez
Department of Physical and Analytical Chemistry,
Faculty of Experimental Sciences, University of Jaén,
Paraje Las Lagunillas, Jaén, Spain

Abstract

Advances in environment, agriculture or medicine raised an increasing demand for control analyses and posed various challenges to analytical chemists such as the need to develop new methods exhibiting as much sensitivity, selectivity, easy automation, low cost of equipment, simplicity, and environmental friendliness as possible. The inception of flow analysis has proved to be an excellent tool for handling solutions and, consequently, performing methods related to wet chemical analysis. In this sense, the main flow methodologies employed in recent years, in addition to conventional flow injection analysis, have been sequential injection analysis, lab-on-valve, and multicommutated flow techniques. Included in the last one, multicommutated flow analysis, multysiringe flow analysis, and multipumping flow systems are the most common. All these computer-controlled techniques have introduced a great number of advantages, mainly a dramatic decrease of the needed sample and reagents volumes, higher repeatability and miniaturization.

This chapter covers a brief description about the fundamentals of the above mentioned methodologies that are normally coupled to different analytical systems, including chromatographic, spectroscopic, and electrochemical ones. A

* E-mail address: anruiz@ujaen.es; Tel: (+34)953-212759; Fax: (+34)953-212940.

critical discussion of the advantages, handicaps and current trends of each methodology will be done, finishing with some applications in the environmental, agri-food and clinical fields.

1. Introduction

At the beginning of the twenty-first century, there was a widespread movement within the chemistry community, with special consciousness of the impact that chemistry was having in both human health and environment. Since then, the chemistry community has focused its efforts in developing less hazardous processes for human health and the environment. This way of doing chemistry was named as 'Green Chemistry', which could be defined as *the use of chemistry techniques and methodologies that reduce or eliminate the use or generation of feedstocks, products, byproducts, solvents, reagents, etc. that are hazardous to human health or the environment*. Analytical Chemistry is one of the branches of chemistry where major efforts are being made to follow the rules of Green Chemistry. These efforts focus on the development and validation of new analytical methods that reduce or eliminate the use and release of hazardous substances in all the steps of the analytical process. This goal may be achieved either by developing new analytical methods as a whole, or by adapting well established analytical methods that are not environmentally-friendly and making them greener. This movement in the Analytical Chemistry, where the operator's safety and environmental consciousness are considered as important as the analytical performance is now known as 'Green Analytical Chemistry'. Different approaches can be followed in order to develop Green Analytical Chemistry and they can be grouped within three main topics:

(1) Consumption of energy. An excess of consumption could cause both environmental and economic impacts. In this sense, Green Analytical Chemistry demands low energy-consumption analytical methods. Thus, if direct analysis can be carried out, this would avoid energy consumption in sample pretreatment and/or preparation. If not, some strategies could be followed in order to achieve this goal. For example, pressurized fluid extraction uses temperatures around 100°C or more for short periods of time (around 10 min) whereas conventional leaching of analytes with organic solvents by means of a hotplate or Soxhlet extraction might use less temperature but larger periods of time (even hours), resulting in an increase of the energy consumed. Supercritical fluid extraction, and microwave- and ultrasound- assisted extraction could also reduce the

time (and energy) of the extraction process. Finally, during the measurement step the consumption of energy is an important issue, since an instrumental analytical technique, which needs electrical energy at least, is usually employed.

(2) Generation of wastes. This is closely related to the chemicals used in the analytical procedure. In general terms, a lower use of chemicals results in a lower generation of wastes. On the one hand, miniaturization should reduce considerably the amounts of wastes produced. On the other hand, recycling and reusing the generated organic wastes could be a good approach, e.g., by means of distillation of the bulk organic waste obtained by different techniques. Another strategy, proposed by some authors, is the on-line waste treatment by means of flow manifolds. The general strategy is to introduce an additional step after the measurement step in order to destroy the toxic species in the effluent. This additional step can be achieved by chemical, photochemical, thermal, or microbiological degradation of the organic species. However, metal ions are non-degradable, and the strategy could consist in chemical or physical adsorption, precipitation or co-precipitation.

(3) Use of chemicals. It should be emphasized that an ideal situation might be that one in which no reagents/solvents are used during any of the steps of the analytical procedure. In this sense, some strategies can be followed, such as the use of:

- *In situ* measurements and/or direct analysis, e.g., by using X-ray fluorescence spectrometry, near infrared spectrometry, and laser spectrometry. However, the reality is quite the opposite of this utopian situation. Therefore, low-toxicity chemicals should be used instead of more hazardous ones and, in any case, the minimum quantity needed should be used.
- Environmentally friendly chemicals, e.g., by replacing hazardous organic solvents involved in sample solution by more innocuous ones (halogenated solvents could be replaced by non-halogenated ones, even ionic liquids; methanol, which is widely used in chromatography, could be replaced by ethanol)
- Alternative extraction techniques, with lower solvent consumption, instead of classical liquid-liquid extraction: solid-phase extraction (SPE), supercritical fluid extraction, pressurized fluid extraction, solid-phase microextraction, and liquid-phase microextraction.

- Automation and miniaturization, reducing the amounts of chemicals; e.g., using flow injection analysis (FIA), sequential injection analysis (SIA), lab-on-valve (LOV), and multicommutation-based flow systems. This chapter will be focused on this topic.

In spite of the progress of some areas of Analytical Chemistry (e.g., electrochemistry) that are developing relevant approaches to address the current challenges of Green Chemistry, spectroscopic methods currently dominate the area of Green Analytical Chemistry. The contribution of flow methods to the development of Green Analytical Chemistry methodologies will be discussed in detail, as they represent a reliable automatic alternative for reducing solvent, sample and reagent consumption; minimizing waste generation; saving time; showing very good sensitivity, selectivity, and precision; and presenting the ability to perform straight forward multi-component analysis.

2. Flow Systems

2.1. FIA

FIA is based on the insertion/injection of a liquid sample into a moving non-segmented carrier stream of a suitable liquid. The injected sample forms a zone, which is transported by the carrier through a coil of tubing to a detector. The detector measures a physical parameter of the sample (absorbance, fluorescence, etc.) that changes continuously as a function of time as the sample passes through the flow cell [1]. This means that the concentration of the species being monitored is continuously changing with time. The carrier may contain a reagent that reacts with the analyte to yield a detectable product, or may consist of an inert solution and in this case the carrier serves as a means of transporting the sample to the detector. Thus, the FIA response curve is a result of two processes, both of kinetic nature, the *physical* process of dispersion of the sample zone within the carrier stream, and the *chemical* process of the formation of chemical species.

In the FIA technique, the basic components are a peristaltic pump to propel the sample and reagents, a series of plastic tubes to carry the solutions, injection valves to introduce constant volumes of sample and reagents in the system, and the detector. A diagram with these components is shown in Figure 1. The main advantages of a FIA system are its repeatability, rapidity and versatility.

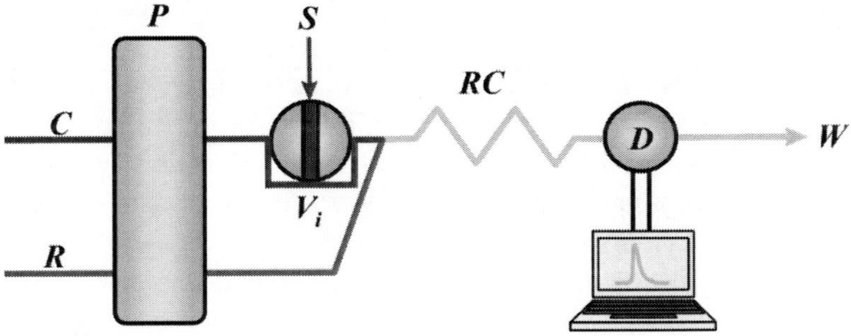

Figure 1. Diagram of a FIA system. P: pump; V: injection valve; RC: reactor (reaction coil); D: detector; C: carrier; S: sample; R: reagent; W: waste.

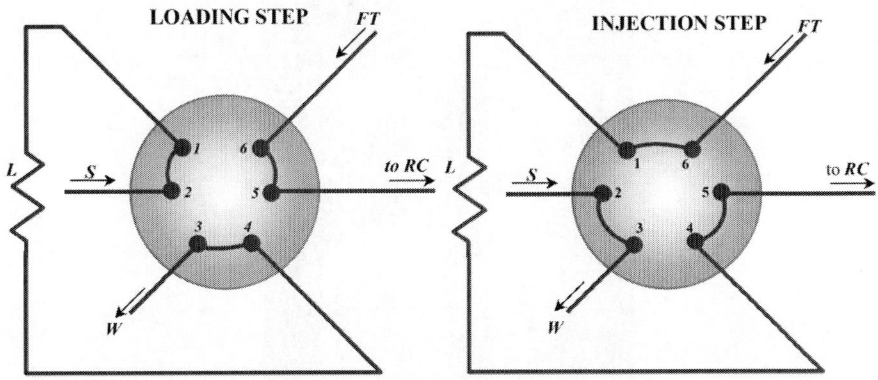

Figure 2. Six-way rotary valve injection procedure.S: sample; R: reagent; FT: carrier; RC: reaction coil; L: valve loop; W: waste.

The injection valve mode is an easy way of inserting well-reproducible sample volumes into the carrier without disturbing its motion. Figure 2 shows a rotary injection valve and its operating mode. The loading and injection steps employed by displacing a movable part between two resting positions are a common feature of these devices. Air bubbles and pressure surges must be avoided during the injection because they will modify the pattern of the flow in FIA system, affecting dispersion and precision.

The valves usually employed are the manually-controlled six-way rotary valves (Figure 3). The sample charge and injection is performed by the movement of a mobile element between two fixed positions, being the sample volume selected by the length of the sample loop and the internal volume of the injection valve.

The tube placed between the injection valve and the detector represents the dispersion coil or the reaction coil and it has a uniform internal diameter (most frequently used being 0.5 or 0.8 mm). It usually is a Teflon tube tightly wound in the form of a coil to promote mixing. Reactors filled with chemically inactive glass beads have been used to improve the mixing without increasing the dispersion. In these rectors a well-controlled and reproducible dispersion pattern of the sample zone and high sample throughput can be achieved.

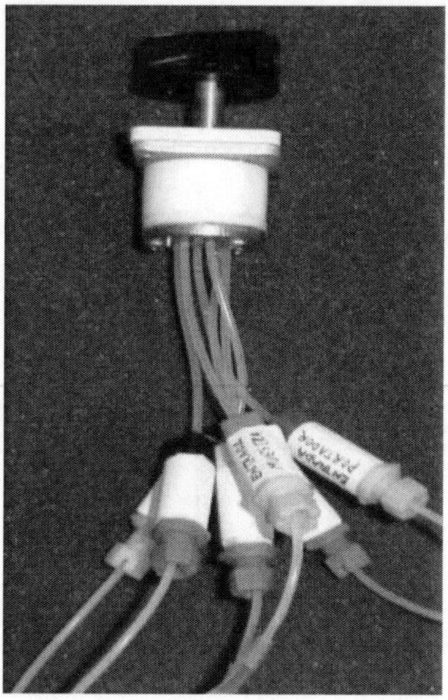

Figure 3. Six-way rotary valve.

2.2. SIA

SIA was introduced as a following generation in the development of flow injection technique [2]. It is based on the same principles as FIA, namely controlled partial dispersion and reproducible sample handling, and it offers different possibilities with a series of advantages and disadvantages in relation to its parent technique. As a robust alternative to FIA, SIA allows implementation of different flow configurations without modifying the manifold. The heart of an SIA manifold is a computer-controlled multi-position selection valve, usually 8-10 ports (Figure 4).

Fluids are manipulated within the manifold by means of a bi-directional pump. A holding coil is placed between the pump and the common port of the multi-position selection valve. The selection ports of the valve are coupled to sample and reagent reservoirs as well as to the detector. The syringe pump draws sample and reagents into the holding coil, mixes them and transports there acting mixture into the flow-through detector. As the zones move towards the detector through the reaction coil, zone dispersion and overlap occur, resulting in the formation of a detectable species that are monitored by the detector.

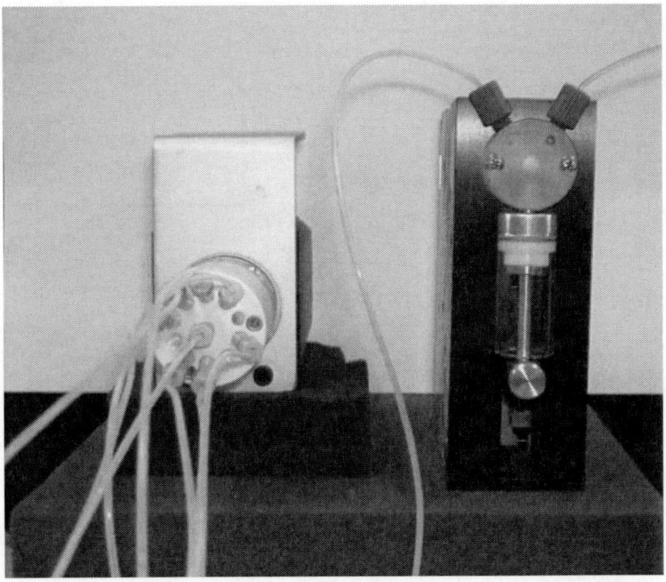

Figure 4. Multi-position selection valve and syringe pump.

Other required components in a SIA system include a suitable flow-cell, tubing/reaction coils and connectors (similar to the ones used in FIA), and a personal computer with an appropriate software (to control the flow direction, the rate and timing of the pump, the position of the multi-port valve; and to collect and process the data). A diagram is shown in Figure 5.

Each measuring cycle begins by switching the multi-position valve to the sample line and aspirating a precise measured volume (few μL) of sample into the holding coil by the pump moving in a backward motion; the pump is stopped during the rotation of the valve to avoid pressure surges. Then, the valve is switched to the reagent line and a precisely measured volume of reagent is drawn into the holding coil. Thus, the sample and reagent solutions are sequentially injected into the holding coil next to one another, hence the name of this technique. A second reagent may be aspirated on the other side of the sample. A suitable volume of carrier solution is aspirated into the syringe pump and used to propel the aspirated sample/reagents. Finally, the valve is switched to the detector port and the pump propels the sequenced zones forward through the reaction coil to the detector.

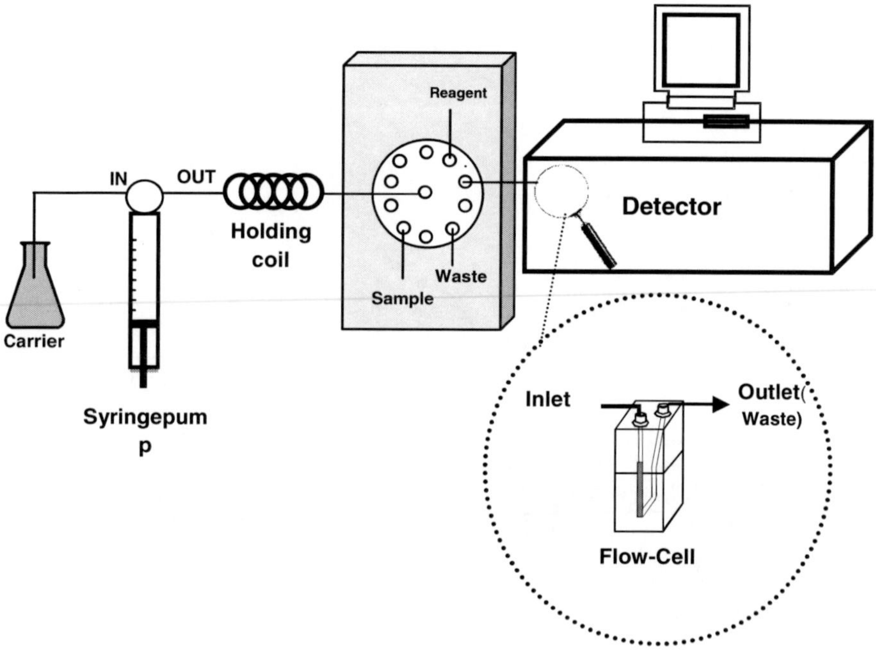

Figure 5. Diagram of a SIA system.

The instrumental simplicity, robustness, ease and efficiency with which hydrodynamic variables can be controlled, and high flexibility and modes maintenance requirements of this modified technique have turned it into a very popular choice with both research and industrial analysis laboratories. On the other hand, SIA has proven to be a technique that can be designed to operate in a multi-parametric way, which is of special interest when considering the design of environmental monitors. However, in spite of these advantages, the major disadvantage of SIA is the lower sample throughput when compared to other flow systems.

2.3. LOV

LOV systems, introduced by J. Ruzicka [3], are considered the third generation in the flow injection technique, after FIA and SIA. LOV was mainly introduced to solve some problems related to reagent-limited assays, handling highly hazardous chemicals, or performing troublesome on-line sample treatments. LOV has made significant progress in miniaturization, automation and integration of on-line sample pre-treatment. It integrates manifold components, including fiber optic-coupled flow-through cell, flow through port, and microcolumns, into a single unit mounted on a conventional multi-position valve (hence the name ''lab-on-valve'').

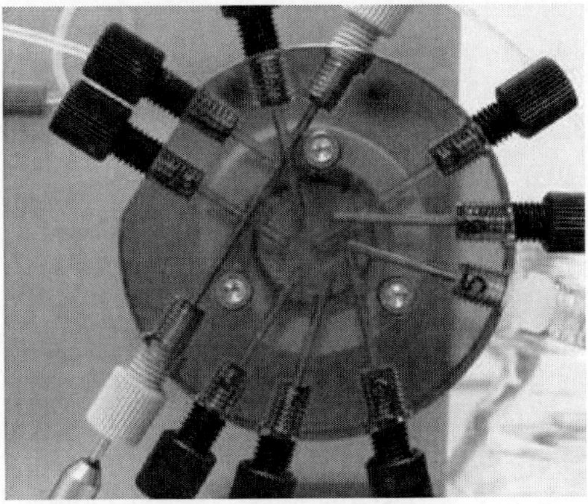

Figure 6. Lab-on-valve (reproduced with permission of FIAlab Instruments, Inc.).

The main advantages of LOV are: a) transparency of flow and optical path; b) versatility of flow-cell design; c) sample introduction by flow-through port; d) miniaturization - compact and robust structure.

The first LOV system consisted in the coupling of SIA and LOV, generating the SI-LOV systems. The syringe pump is still the flow driving device, but the conventional multi-position valve is replaced with the LOV. The introduction of LOV concept achieves downscaling reagent-based analysis to microliter levels. In contrast to other microfluidic devices, the volume of the sample path within the LOV is minimized by reducing the channel length, while channel diameter is kept wide in order to prevent system clogging by matrix material found in real samples. A multi-purpose flow cell is permanently incorporated in the LOV, allowing several miniaturized fluidic operations to be performed, such as sample dilution and reagent addition/mixing. The cell is furnished with optical fibers, communicating with an external light source and a detection device, and different detection techniques can be used, changing the measurement angle of the fibers for absorbance or fluorescence recordings.

LOV has proved especially successful in the development of flow systems in which on-line solid-phase extraction (SPE) is performed, or for the implementation of Bead Injection (BI), generating the SI-LOV-BI methodology. Some examples of the use of SI-LOV and SI-LOV-BI systems will be described in section 4 of the chapter.

2.4. Multicommutation

Multicommutation is based on the use of discrete commutation devices to build up dynamic manifolds that can be easily reconfigured using appropriate software, so that each analytical step of the procedure can be implemented independently. Usually, one commutation device is employed to manage each solution in an automated fashion and only the reagent volume necessary for sample analysis is introduced. Micro-volumes of samples and reagents can be sequentially inserted into the reaction coil of a single line manifold by electronic control of the valve solenoid, avoiding excessive use of chemicals.

The main difference between multicommutation and the other flow methodologies is that insertion volumes are replaced with insertion times [4]. Among the multicommuted flow techniques, multicommuted flow analysis (MCFIA), multipumping flow systems (MPFS), and multisyringe flow analysis (MSFIA) are the most widely employed ones. They combine the advantages given by FIA such a high injection throughput with those given by SIA such as a minimal reagent consumption and waste generation.

MCFIA systems are typically constituted by a peristaltic pump and a set of three-way solenoid valves, automatically controlled by appropriate software, which can be arranged creating a flow network. Each valve can adopt two positions, being the whole system assimilated to an electronic circuit with a variable number of active nodes. MPFS systems are based on the use of a series of individual solenoid micro-pumps acting as liquid propelling units. These micro-pumps are responsible for sample-reagent introduction and manifold commutation. And finally, MSFIA relies on a device designated by a multisyringe burette which is a multiple channel piston pump, driven by a single motor of a usual automatic burette and controlled by computer software through a serial port. All these methodologies will be explained in more detail.

2.4.1. MCFIA

MCFIA technique was devised by B.F. Reis et al. [5], being the key element in these systems the solenoid valve, usually 3-way solenoid valves. These valves act as a switch between two different positions, "OFF" and "ON", remaining permanently connected two of the three valve ports. The switching procedure of the three-way solenoid valve is shown in Figure 7. There are two main configurations for the design of a MCFIA system: aspiration of the solutions or propulsion of the solutions. On the one hand, when the solutions are aspirated (Figure 7(A)), one valve can be used for controlling the flow of two different solutions. As it can be observed, one port is always opened, and it can be communicated with one of the other two ports by means of the up and down movement of the solenoid; therefore, the solution required is selected by activating or deactivating the 3-way solenoid valve.

On the other hand, when the liquids are propelled, only one solution is controlled by each individual solenoid valve. In Figure 7(B), the scheme for the carrier solution flow is shown. When the valve is deactivated, the carrier flows towards the detector; however, when the solenoid valve receives the electric pulse, the solenoid moves up and the carrier is recycled to its corresponding vessel. In the case of sample and reagent solutions, the recycling and detector ports would be exchanged. Hence, the sample/reagent solution would be inserted in the system only when the valve is activated.

The sample volume inserted is directly related with the time of the electric pulse, and it can be changed by modifying the insertion sequence profile. It is possible to insert the sample volume in one or several aliquots, therefore mixing the sample with the carrier solution or with reagents solutions. The volume of sample spent is only that one aspirated during the time the valve remains "ON". As a result, a high saving in terms of consumption of sample and reagents solutions is obtained by employing this automated methodology. Peristaltic pumps, automatic burettes or syringe pumps have been employed for propelling the liquids in these systems.

2.3.2. MPFS

These systems, developed in 2002 by two research groups at the Pharmacy Faculty of the University of Porto (Portugal) and the Piracicaba CENA (Brazil), are based on the combined use of solenoid valves and pumps, leading to the so called solenoid piston pumps or micro-pumps [6]. The MPFS manifold is made up of a series of solenoid micro-pumps, as many as the solutions required by the analytical method. The micro-pumps operate in the form of pulses in a monodirectional way. Each pulse produces a piston stroke of the pump, which propels a preset volume of liquid (depending on the model of the micro-pump). The number of pulses defines the total injected volume and the flow-rate is determined by the stroke frequency. As a result, these systems are similar to a MCFIA system, where the peristaltic pump is replaced by the solenoid pumps, which also act as the solenoid valves themselves (opening or closing the flow of liquids).

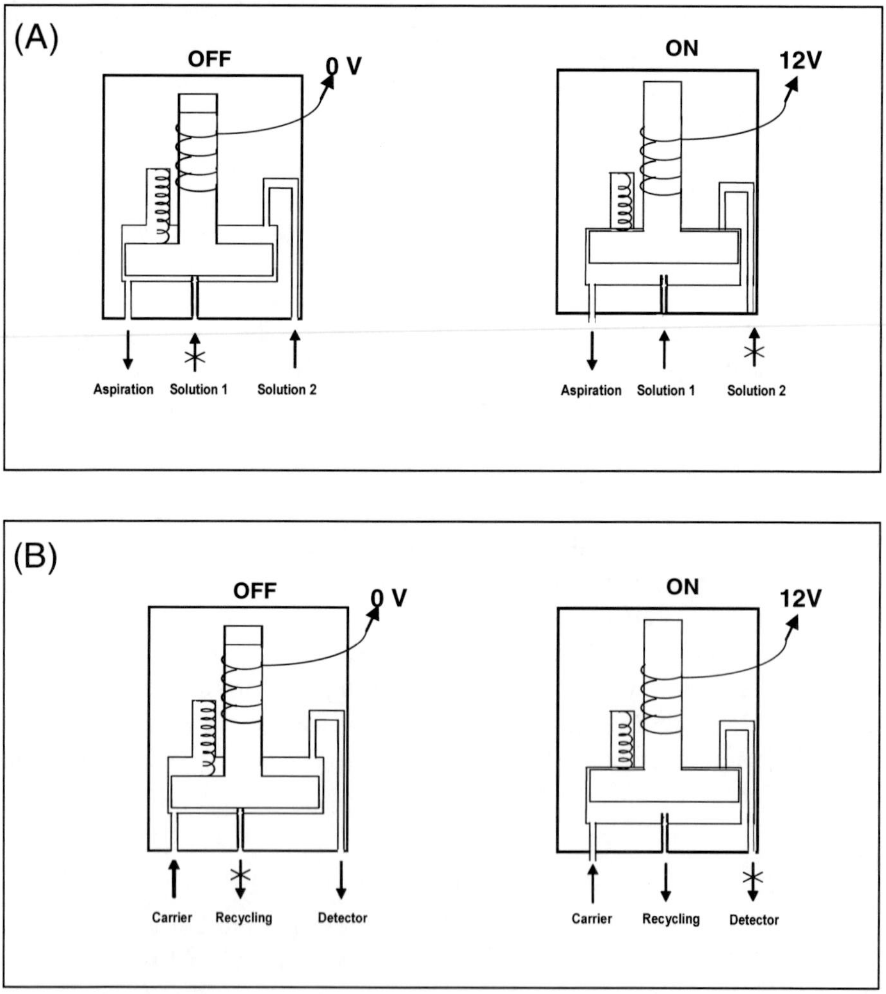

Figure 7. Procedure of the three-way solenoid valve.

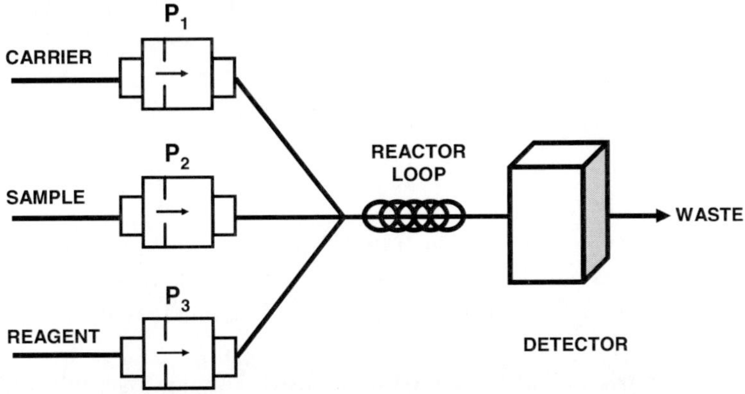

Figure 8. MPFS manifold.

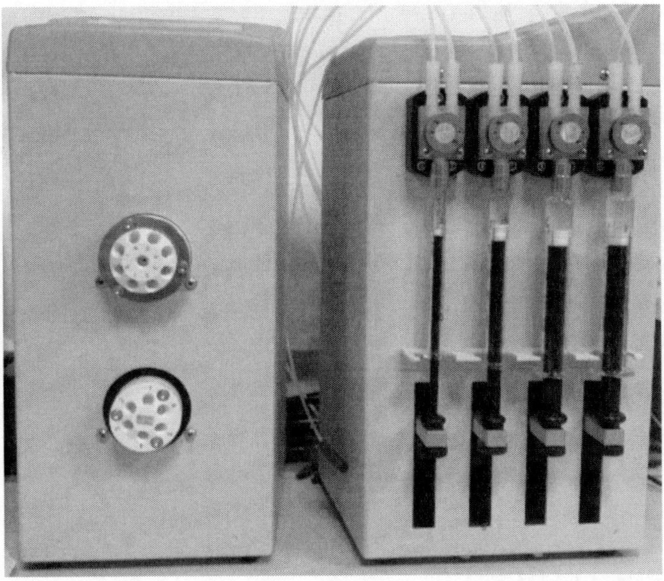

Figure 9. MSFIA system.

The principal advantages of a MPFS are its high flexibility, easy configuration, robustness, and low cost (as explained, the solenoid pump operates as both the valve and liquid propeller). In addition, considering that each micro-pump acts individually in the propulsion of the fluids, distinct sampling techniques (such as the single sample volume, binary sampling or merging zones) can be explored in a versatile and independent way. The better mixing of reagents and sample due to the pulse mode provides higher peaks when compared with other flow techniques. A typical and simple manifold is depicted in Figure 8, in which only 3 solenoid pumps are used, being responsible for the individual handling of three different solutions: carrier, reagent and sample.

2.3.3. MSFIA

MSFIA was developed by Cerdá et al. (Department of Analytical Chemistry, University of Balears Islands, Spain) together with the firm Crison (Alella, Barcelona, Spain) with the aim of combining the advantages of both SIA and MCFIA [7].

The device consists of a conventional automatic titration burette adapted in a way that the motor can simultaneously move the pistons of four syringes, avoiding the need of using separate burettes in parallel (Figure 9). This movement is accomplished by using a horizontal metal bar, moved by the motor of the burette, which accommodates the four syringes. In addition, each syringe has a three-way solenoid valve placed at its head, which is used to select the direction of the liquid flow. As the burette motor moves the four syringes at the same time, different flow-rates can be achieved in each channel by modifying the dimensions of the syringes. One or two multi-position valves are usually employed in MSFIA systems.

The main advantages of MSFIA are: robustness of the system; possibility of performing direct sample and reagent confluence (in classical SIA there is only one syringe); and its capacity to work under moderate backpressure. The propulsion system used opens up new possibilities, combining the multichannel operation of peristaltic pumps with the constant, pulseless and exactly known volume delivery achieved by piston pumps. However, MSFIA has the same disadvantage as for piston pumps that the forward movement must be stopped to reload the syringes, decreasing the sampling frequency.

3. Detection Techniques

Although every flow system offers inherent outstanding advantages, they have to be complemented by selective and sensitive detection techniques, additional separation devices and clean-up steps when analyzing complex matrices or determining more than one analyte. In this section, the detection techniques more frequently coupled to flow systems are described.

3.1. Optical Detection

3.1.1. UV-Visible Spectroscopy

UV-Visible spectroscopy has been traditionally the most frequently used detection technique in analytical laboratories due to its high flexibility for adaptation to a wide variety of analytical problems and the low-cost of equipments. However, not many analytical methods rely on the intrinsic absorbance of the analytes, as their intrinsic light absorption occurs at the UV region, where other co-existing compounds also absorb, interfering in the analytical measurements. As a result, UV region does not usually present the required selectivity for conventional spectrophotometric analysis. However, derivatization has been frequently described, especially when determining inorganic species.

3.1.2. Luminescence

The use of luminescence detection techniques has gained considerable ground in analytical chemistry due to the high sensitivity and selectivity that can be achieved. Moreover, the combination of the FI method with luminescence detection has become increasingly important in environmental analyses because of its wide linear dynamic range, reproducibility, simplicity, rapidity and feasibility.

However, not all luminescence techniques are selective enough, and different strategies can be employed in order to overcome some limitations, such as spectral overlapping, scattered light, quenching processes, or background signal. The techniques more widely employed are:

- *Fluorescence*. The intrinsic fluorescence of the analyte has been used for the determination of several compounds. Nevertheless, as there is a small number of compounds that exhibit native fluorescence, derivatization reactions have been employed to increase the number of applications. One novel approach is the so called photochemically-induced fluorescence (PIF). It consists in the generation of fluorophores from non-fluorescent analytes by on-line UV irradiation and presents inherent advantages over ordinary chemical reactions such as quicker reaction rate, fewer chemicals involved and smaller dilution factor.
- *Phosphorescence*. When phosphorescence is selected as detection technique, a common aspect is the need for some form of molecular immobilization

and/or protection in order to minimize non-radiative decay of luminophores, collisions with solvent, or the possibility of photochemical reaction. Under a specific condition phosphorescence from a chemical component can be observed even at room temperature. This long-lived emission is called as room temperature phosphorescence (RTP). When comparing the RTP method with fluorescence, the RTP method has some advantages: the short-lived background luminescence or scattered light can be easily discriminated because of the long-lived emission of the analyte species, and very good separation of the maximum wavelength of phosphorescence from that of the excitation spectra due to Stoke's shifts is achieved.

- *Chemiluminiscence*. This is a phenomenon involving the emission of light (usually in the visible or infrared region) as a result of a chemical reaction. Essentially, the process by which the luminescence is produced is identical to that for fluorescence except that no excitation light source is necessary. The analytical interest of chemiluminiscence arises from the ability to produce fluorescent molecules with no prior irradiation, thereby avoiding various problems derived from light scatter, unselective excitation or light source instability.
- *Lanthanide-sensitized luminescence*. The trivalent cations of the lanthanides have photoluminescent properties that are favorable for several kinds of applications. However, it is difficult to generate this luminescence by direct excitation of the lanthanide ion, because of the ions' poor ability to absorb light. Nevertheless, they possess unusual spectroscopic properties when chelated with appropriate organic ligands. These chelates display a well-defined luminescence characterized by narrow and highly structured emission bands, large difference between absorption and emission wavelengths (Stokes's shifts) and long excited-state lifetimes. As a result, these chelates have been widely used with chemical and biological applications, mainly for the determination of organic analytes. This detection technique is called lanthanide-sensitized luminescence. Once the chelate is formed, the lanthanide ion luminescence originates from an intramolecular energy transfer through the excited state of the ligand (organic analyte) to the emitting level of the ion. Tb (III) and Eu (III) are generally preferred and used in practical applications because of their longer decay times and intense luminescence intensities.

3.1.3. Vibrational Spectroscopy

Although vibrational spectroscopic detection does not present the sensitivity observed in most of the systems developed employing luminescence detection, it provides structural information. The construction of systems based on Fourier-transform infrared spectroscopy (FTIR) detection is complex because the path length of the flow cell must be kept very short due to the high absorption of water in infrared range. Thus, the extraction of the analyte from the aqueous solution seems to be a valuable alternative, as this procedure avoids the interference of water. The use of near infrared detection allows working with glass cells and longer path lengths, but the superposition of overtones and combination bands in this spectral region causes a lower structural selectivity. In this context, Raman spectroscopy appears as a very interesting alternative since water is a weak Raman scatterer and conventional glass flow-cells can be used to construct flow systems.

3.1.4. Atomic Spectroscopy

Although the analysis of metals can be carried out by UV-Vis spectroscopy or luminescence using derivatization reactions, their analysis is usually performed by means of atomic absorption/emission spectroscopy. The techniques more widely employed are:

- *Atomic Absorption Spectroscopy (AAS).* Light of a wavelength characteristic of the element(s) of interest is shone through this atomic vapour. Then, some of this light is absorbed by the atoms of that element(s), being the amount of the absorbed light measured and used to determine the concentration of that element(s) in the sample. The most common techniques of AAS are flame atomic absorption spectrometry (FAAS) and graphite furnace atomic absorption spectrometry (GFAAS). In FAAS a flame is used to vaporize and atomize the sample introduced through a nebulizer, whereas in GFAAS the sample is deposited into a graphite tube inside a furnace, and is vaporized and atomized when the graphite tube is heated. FAAS presents less sensitivity than GFAAS, and a higher amount of sample is also required in FAAS. However, FAAS is the least expensive detection technique, and requires less of a skill level for an analyst than GFAAS. In addition, analyses performed using FAAS can be much quicker than those performed using the more time-consuming GFAAS. However, both techniques present the inconvenience of the non-multi-element analysis, therefore obtaining low sample-throughputs when high numbers of analytes need to be determined. In those cases, inductively coupled plasma with optical or mass spectrometry detection is normally used.
- *Atomic Emission Spectroscopy (AES).* Flame emission spectroscopy, and spark and arc AES are used, although inductively coupled plasma-optical emission spectroscopy (ICP-OES) and atomic fluorescence spectroscopy

(AFS) are probably the most important detection techniques. In flame emission spectroscopy, a sample of a material is brought into the flame as either a gas, sprayed solution, or directly inserted into the flame by use of a small loop of wire. The heat from the flame evaporates the solvent and breaks chemical bonds to create free atoms. The thermal energy also excites the atoms into excited electronic states that subsequently emit light at the characteristic wavelength when they return to the ground electronic state. Spark and arc AES are used for the analysis of metals in solid samples. An electric spark or spark is passed through the sample, heating it to a high temperature to excite the atoms within it. The excited analyte atoms emit light at characteristic wavelengths.

ICP-OES, also called ICP-AES, has been widely used for the determination of metals in different samples in the last years. ICP-OES utilizes an argon plasma to vaporize, atomize, excite and ionize the elemental species in a sample. Once the atoms or ions are in their excited states, they can decay to lower states through radiative (emission) energy transitions. The intensity of the light emitted at specific wavelengths (characteristic of the element) is used to measure the content of the metals in the sample. ICP-OES allows sensitive multi-element analysis because characteristic emission spectra may be generated for a range of elements under the same excitation conditions. In addition, ICP-OES also presents the possibility of choosing from several different emission wavelengths of an element.

AFS is based on the absorption of radiation of specific wavelengths by an atomic vapour, followed by detection of radiationally deactivated states *via* emission in a direction (typically) orthogonal to the excitation source. Both absorption and subsequent atomic emission processes occur at wavelengths that are characteristic of the atomic species present. AFS is a very sensitive and selective method for the determination of some toxic metals such as As, Cd, Hg, or Pb.

3.2. Electrochemical Detection

In comparison to other instrumental methods of chemical analysis, electroanalytical instrumentation is relatively easy to miniaturize. The main trends in modern electroanalysis include development of chemical and biochemical systems based on progress in chemical and biochemical methods of molecular recognition and development of measuring devices of a large integration scale, including sensor arrays. This is accompanied by the use of new materials, including nanomaterials and nanostructures, as long as the introduction of flow systems, which is the aspect considered in this review.

Most electrochemical detectors, such as amperometric and potentiometric detectors, are surface detectors. They respond to substances that are either oxidizable or reducible and the electrical output results from an electron flow caused by the chemical reaction that takes place at the surface of the electrodes. Successful operation of a surface detector requires reproducible radial concentration distribution. There are several types of flow detection cells, each type being characterized by parameters such as the length, diameter, and shape of its detection channel, which determine the laminar character of the liquid flow under the given experimental conditions and the predominant mode of the mass transport within the cell.

Electrochemical detectors present special features for flow analysis, such as: (1) the flowing liquid continuously cleans the surface of the indicator or working electrode; consequently, the electrochemical and/or mechanical regeneration of the working electrode surface with intensive washing, solution or solvent switching, potential cycling, etc. are generally less crucial when compared to batch electroanalytical techniques; (2) the continuously streaming carrier solution removes reaction products and impurities leached from the electrode; (3) the convective transport of an analyte and/or a reactant reduces response time and improves the detection limit compared to batch-type measurements with pure diffusion transport; and (4) differences in the response rate for the primary and interfering ions, in case of potentiometric detectors, can improve selectivity.

3.3. Mass Spectrometry Detection

Inductively coupled plasma-mass spectrometry (ICP-MS) is one of the most common used detection techniques for the analysis of traces of metals in a wide variety of samples. The high temperatures obtained in an ICP, up to 8000-10000 K, make it the most versatile atomizer and element ionizer available, as no material can withstand that temperature. ICP-MS uses an argon plasma to vaporize, atomize and ionize the elemental species in the sample solution, in a same way than ICP-OES. However, in ICP-MS, a mass spectral detection is used instead of a wavelength-based detection. The most important advantages of ICP-MS are its high sensitivity; multi-element quantitation; relative salt tolerance; compound-independent response, leading to species-unspecific measurements; large dynamic range; and high quantitation accuracy. ICP quadrupole MS (ICP-Q-MS) presents some disadvantages compared to other detection techniques, mainly the presence of spectral interferences from the matrix constituents, and its high price compared to other equipments. To eliminate the spectral interferences, different alternatives are used: (1) sample pre-treatments; (2) inter-element correction algorithms; (3) measurement of less abundant isotopes (such as ^{57}Fe instead of ^{56}Fe); or (4) collision and reaction cell technologies.

GC-MS and HPLC-MS techniques are used in a high percentage of the currently developed analytical methods, due to their outstanding selectivity and sensitivity, and the possibility of the quantification of numerous analytes in the same analytical run. Flow methodologies are coupled to chromatographic techniques with MS detection mainly with the intention of performing on-line sample treatments previous to the determination of the target compounds.

4. Applications

All the flow methodologies and detection techniques, previously described, can be coupled so increasing the automation and repeatability of the systems, and decreasing the consumption of reagents and the time required per analysis. However, in complex samples, more than half of the analysis time can be spent on the sample preparation steps, being the quality of these steps a key factor in the success of the whole analysis, even more important than the detection technique itself. Traditionally, sample preparation has been performed off-line, but automated flow methodologies can be used in sample pre-treatment procedures with different purposes, such as extraction, pre-concentration, clean-up, derivatization, etc. These on-line sample pre-treatments show outstanding advantages in all areas of analytical chemistry, as have been demonstrated in the last few years. Several interesting examples of the usefulness of the different flow methodologies will be discussed in this section to show their potentiality in modern analytical methods in the environmental, alimentary, and clinical fields.

A flow injection on-line dialysis (FID) was developed for sample pre-treatment prior to the simultaneous determination of several food additives (acesulfame-K, saccharin, caffeine, benzoic acid, and sorbic acid) by HPLC [8] (*Example 1*). The FID-HPLC manifold is shown in Figure 10.

The procedure utilized in the FID-HPLC system was as follows. Firstly, eluent was pumped through the HPLC column to clean the system. The donor (sucrose, P_1) and acceptor (phosphate buffer, P_2) solutions of the FID system were flowed to fill all tubings and channels. For the first injection, a mixed standard or sample solution at V_1 was injected into the donor stream and then it was propelled to the DC, where small solute molecules in the donor solution were dialyzed through the dialysis membrane into the acceptor solution. In order to enhance the sensitivity, the donor stream was continuously flowed while the acceptor stream was stopped during the dialysis period. The dialysate zone containing food additives was filled into the sample loop of the HPLC valve (V_2) and was injected into the HPLC system and analyzed under normal HPLC conditions. While the dialysate of the first injection was injected into the HPLC system, the second injection was loaded into the sample loop of the FID valve (V_1). Hence, when the chromatographic separation of the first injection was ended, the dialysate of the second injection was injected into the HPLC system, and the third injection was loaded into the FID system. In this way, the FIA manifold was useful not only for the sample treatment, but also to increase the sample throughput of the whole system.

The recent development of monolithic columns for HPLC has allowed efficient separations to be performed at far lower pressure, offering the opportunity to perform separations in some flow-analysis manifolds that would not withstand the back-pressure from conventional packed columns.

In this sense, FIA and SIA methodologies have been used to develop different analytical methods. For instance, a FIA-HPLC system has been described for the determination of six opiate alkaloids (morphine, pseudomorphine, codeine, oripavine, ethylmorphine, and thebaine) and four biogenic amines (vanilmandelic acid, serotonin, 5-hydroxyindole-3-acetic acid, and homovanillic acid) in human urine, using CL detection [9]. Some examples of the use of SIA-HPLC, called sequential injection chromatography, include the quantitation of vitamins A-acetate, D2, D3, and E-acetate [10] and the determination of the pesticides fenoxycarb and permethrin [11], in both cases employing UV detection.

A novel SIA manifold has been recently described for the automation of liquid-liquid extraction, using two multiposition valves in order to separate the extraction and determination units. This system is called double-valve-SIA (DV-SIA), and has been used for the spectrophotometric determination of thiocyanate in saliva samples after extraction with amylacetate [12] (*Example 2*). The manifold is shown in Figure 11. Briefly, the procedure was divided in the following steps: a) water, used as carrier, was aspirated in the extraction part; b) an air plug was aspirated into the holding coil, followed by astra phloxine solution, buffer solution, sample solution, and amylacetate solution; c) all aspirated zones were delivered into the extraction cell, and the air was used for efficient mixing of the solutions and to speed up the extraction of the ion associate into the organic phase; d) the phases were self-separated due to their different densities, leaving the organic phase with the extracted compound in the upper part; e) the aspiration of amylacetate, used as a carrier, was performed in the detection part, and the aspiration and pumping to the detector of the extracted ion associate was carried out, followed by its spectrophotometric measurement in the detector.

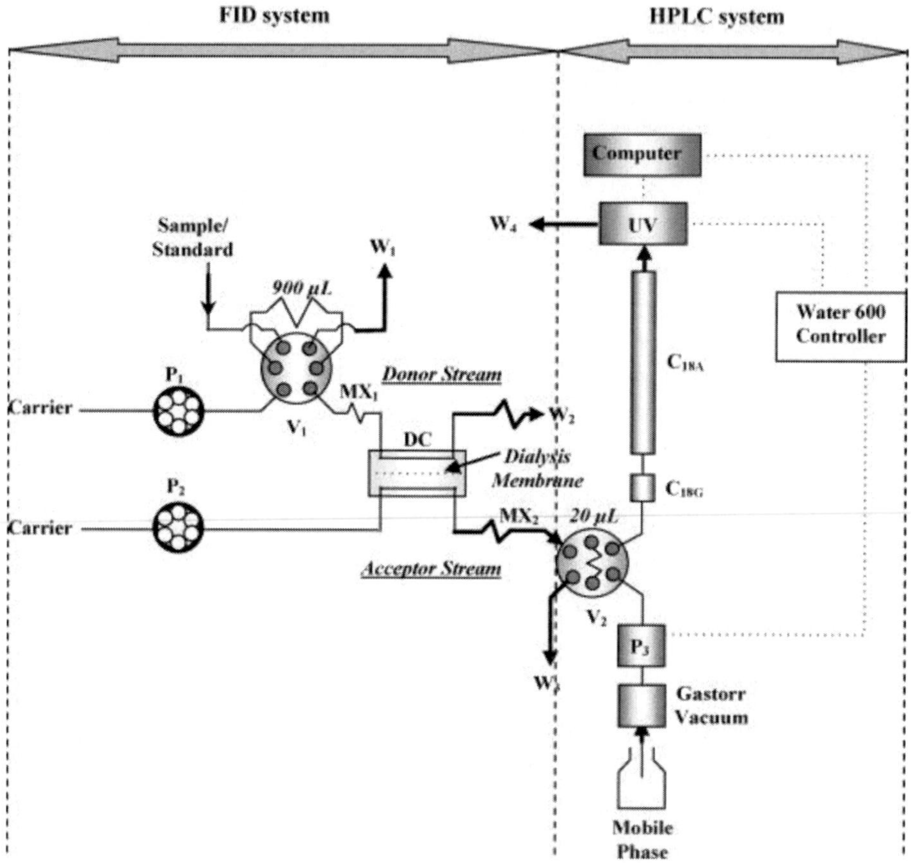

Figure 10. Manifold of the FID–HPLC system. P_1 and P_2: peristaltic pumps, P_3: HPLC pump, V_1: manual rotary injection valve, V_2: HPLC injection valve, DC: dialysis cell, MX_1, MX_2: mixing coils, C_{18A}, C_{18G}: C_{18} analytical and guard columns, UV: photodiode array detector, W_i: wastes. (Reprinted from: Kritsunankul, O. & Jakmunee, J. (2011). Simultaneous determination of some food additives in soft drinks and other liquid foods by flow injection on-line dialysis coupled to high performance liquid chromatography. *Talanta*, 84, 1342-1349, with permission from Elsevier).

An analytical procedure for multiple standard additions of arsenic species using SIA was proposed for their quantification in seafood extracts [13] (*Example 3*). SIA manifold (Figure 12) presented flexibility for generating multiple species standards at the ng mL^{-1} concentration level by adding different volumes of

As(III), As(V), monomethylarsonic (MMA) and dimethylarsinic (DMA) to the sample. The mixed sample plus standard solutions were delivered from SIA to fill the HPLC injection loop. Subsequently, As species were separated by HPLC and analyzed by AFS. In this work, SIA was not only useful for the standard addition method employed, but also for reducing the hazardous As wastes residues. The manifold used in this system is presented in Figure 12, where the syringe pump and multiposition valve that comprised the SIA system are shown.

A SIA manifold (Figure 13) has been used previously to ICP-MS detection for matrix removal and preconcentration (using an ion-exchange resin) and applied to the determination of Al, As, Co, Cu, Mn, Mo, Ni, Pb and V in seawater [14] (*Example 4*). The preconcentration manifold was configured in such a way that the 2% HNO_3 solution was pumped continuously to the nebulizer via the peristaltic pump of the instrument during the loading and preconcentration of analytes. Briefly, the SIA program consisted in the following steps: a) loading of sample solution into the column; b) elution of alkaline metals using an ammonium acetate solution; c) by-pass valve in elution position for the elution of the analytes with 2M HNO_3 and determination by ICP-MS; d) conditioning of the column with ammonium acetate, being the system prepared for the next sample solution.

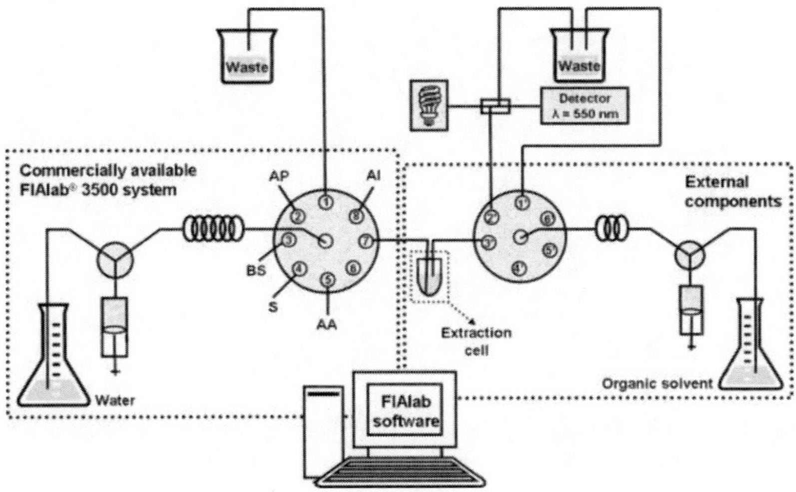

Figure 11. Schematic view of the DV-SIA manifold for online liquid–liquid extraction. AP: Astra Phloxine; BS: buffer solution; S: sample; AA: amylacetate; AI: air input. DV-SIA manifold (Reprinted from: Acebal, C.C., Sklenářová, H., Škrlíková, J., Šrámková, I., Andruch, V., Balogh, I.S. & Solich, P. (2012). Application of DV-SIA manifold for determination of thiocyanate ions in human saliva samples. *Talanta*, 96, 107-112, with permission from Elsevier).

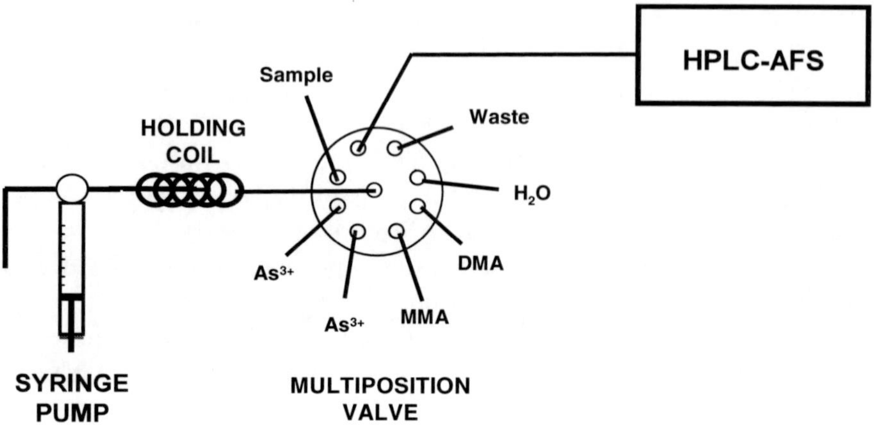

Figure 12. Scheme of the SIA manifold used previously to the HPLC-AFS system.

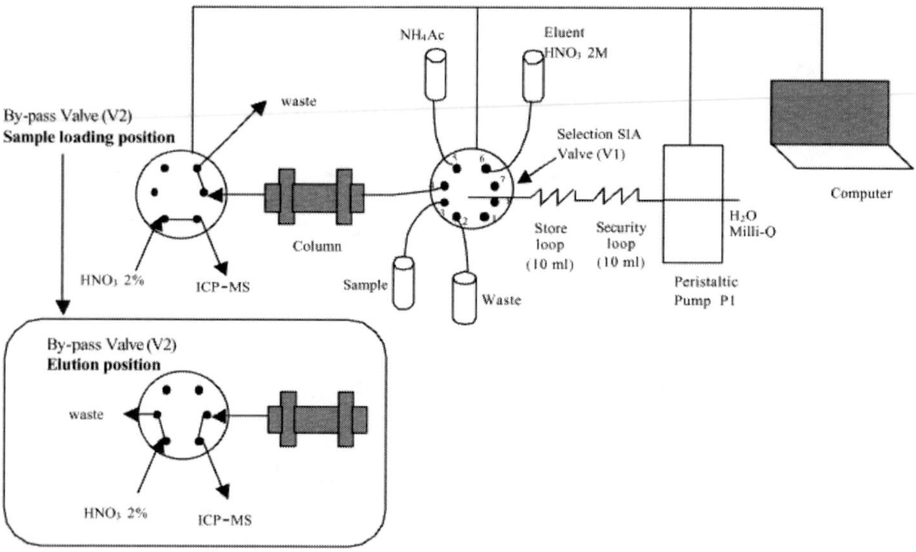

Figure 13. Schematic diagram of the SIA preconcentration system. (Reprinted from: Jiménez, M.S., Velarte, R. & Castillo, J.R. (2002). Performance of different preconcentration columns used in sequential injection analysis and inductively coupled plasma-mass spectrometry for multielemental determination in seawater. *Spectrochimica Acta Part B: Atomic Spectroscopy*, 57, 391-402, with permission from Elsevier).

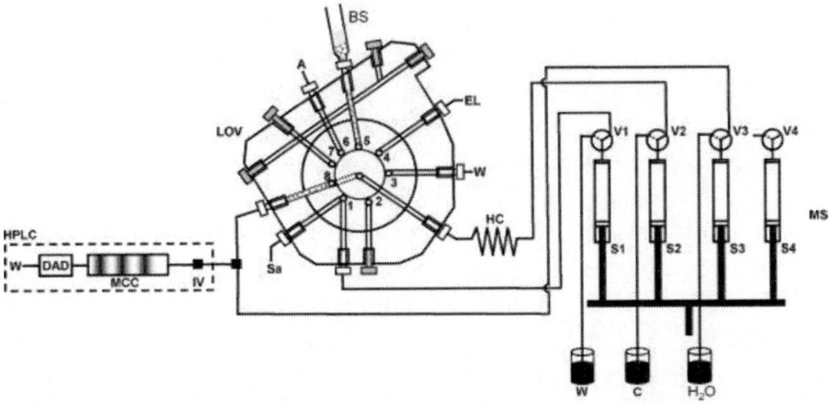

Figure 14. Representation of the MSFI-BI-LOV system hyphenated to HPLC for the determination of UV filters. MS: multisyringe; V_i: three-way commutation valve; S_i: syringe; C: carrier solution; EL: eluent; Sa: sample or standard solution; BS: bead suspension; W: waste; HC: holding coil; A: air; MCC: monolithic chromatographic column; IV: injection valve; DAD: diode array detector. Closed ports at LOV module are represented by grey fittings. (Reprinted from: Oliveira, H.M., Segundo, M.A., Lima, J.L.F.C., Miró, M. & Cerdá, V. (2012). On-line renewable solid-phase extraction hyphenated to liquid chromatography for the determination of UV filters using bead injection and multisyringe-lab-on-valve-approach. *Journal of Chromatography A*, 1217, 3575-3582, with permission from Elsevier).

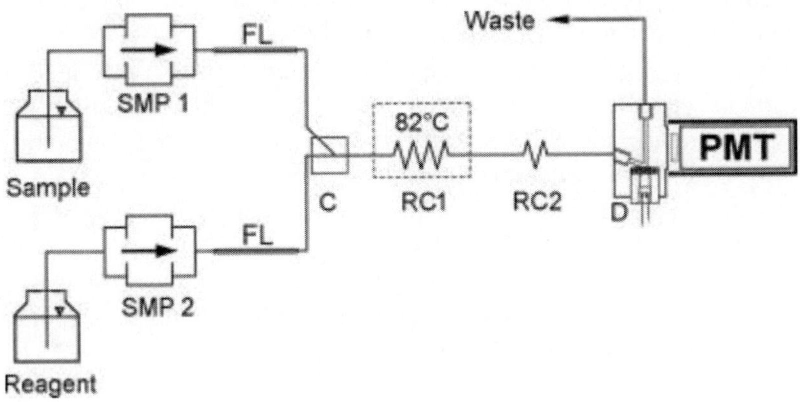

Figure 15. Schematic drawing of the MPFS. (Reprinted from: Horstkotte, B., Duarte, C.M. & Cerdá, V. (2011). A miniature and field-applicable multipumping flow analyzer for ammonium monitoring in seawater with fluorescence detection. *Talanta*, 85, 380-385, with permission from Elsevier).

The use of micro-SIA coupled to LOV (μSIA-LOV) permits the miniaturization and portability of the instrumentation, aspect that is especially useful for the in-situ analysis of samples. For instance, this system has been proposed for the onboard quantitation of Zn in seawater, using fluorescence detection after proper derivatization [15].

An automated sample pretreatment/detection method was proposed for the determination of UV filters (benzophenone-3, ethylhexylmetoxycinnamate, butylmethoxydibenzoylmethane and homosalate) in environmental waters (*Example 5*). The methodology comprised in-line SPE of the target analytes using a MSFIA-BI-LOV system, followed by determination by HPLC [16] (Figure 14). A brief description of the method follows: a) p*acking and conditioning of the extraction column:* *a*fter filling S_2 with carrier, and the HC with 95% (v/v) MeOH/H_2O, the volume of bead suspension was collected through port 5. The sorbent was packed within the cavities of LOV by forward flow pumping of the beads plus methanolic solution and carrier solution; b) then, the s*ample was aspirated and* perfused through the sorbent, being the target analytes retained. 3 volumes of sample were introduced in each analytical run to increase the sensitivity of the system. The sample loading step was finished by removing the non-retained species with carrier solution; c) the elution of the analytes was performed by propelling the required volume of the methanolic solution through the sorbent bed, and the mixture that filled the loop of the injection valve was injected into the chromatographic system; d) *t*he packed SPE microcolumn was wetted with the methanolic solution and the used beads were aspirated back into HC and disposed to waste through port 3 of the LOV.

Finally, as last example (*Example 6*), a miniaturized flow analyzer for ammonium in seawater was constructed using solenoid micropumps (SMP) [17] (Figure 15). For each analytical run, the sample and reagent (sodium tetraborate, ortho-phthaldialdehyde, and sodium sulfite) solutions were injected in the system by simultaneous pulses, they were mixed at the confluence C and the mixture was driven into the heated reaction coil (RC_1). After the reaction product isoindol-1-sulfanat was formed, the solution was pushed through RC2 to cool down the mixture, and it was finally driven through the detection cell for fluorescence quantification.

Conclusions and Trends

Green analytical chemistry has shown consistent growth, pushed by the increased conscience of researchers and analysts in relation to the wastes generated in analytical laboratories. Independently of their analytical performance, methods that are not environmentally friendly will become unacceptable in the near future. The development of automated flow methodologies has contributed to minimize the impact of analytical activities, but its potential has not been fully exploited yet, opening a promising research field.

All the flow methodologies described in this chapter present intrinsic favorable characteristics, as previously stated. However, each one presents specific advantages when compared to the others. For instance, the introduction of SIA, multicommutation and LOV methodologies has provided additional automation when compared to FIA, due to the complete absence of human intervention during each analytical run. In FIA, an operator has to manually switch the rotary valves to inject the sample or reagents solutions, but this step is automatically controlled by appropriate software in recent automated methods.

In general, the coupling of the described methodologies to different detection techniques has provided the laboratories with favorable tools in order to develop routine methods of analysis. However, in recent years, the trends in the development of new analytical methods focus in the automation of the whole analytical process, including sample treatment. Some examples of on-line sample treatments have been described to demonstrate the usefulness of flow methods in different analytical procedures. Although an increasing number of scientific papers is being published year after year in this area, more research is still required. In our opinion, new developments in this area will focus in the continuous miniaturization of the flow systems, and progressive automation of the diverse sample treatments currently available.

References

[1] Ruzìcka, J. & Hansen, E.H. (1975). Flow Injection Analyses. Part I. A New Concept of Fast Continuous Flow Analysis. *Analytica Chimica Acta*, 78, 145-157.

[2] Ruzicka, J. & Marshall, G. D. (1990). Sequential Injection: A New Concept for Chemical Sensors, Process Analysis and Laboratory Assays. *Analytica Chimica Acta*, 237, 329-343.

[3] Ruzicka, J. (2000). Lab-on Valve: Universal Microflow Analyzer Based on Sequential and Bead Injection. *Analyst*, 125, 1053-1060.

[4] Llorent-Martínez, E. J., Ortega-Barrales, P., Fernández-de Córdova, M. L. & Ruiz-Medina, A. (2010). Multicommutation in Flow Systems: A Useful

Tool for Pharmaceutical and Clinical Analysis. *Current Pharmaceutical Analysis*, 6, 53-65.

[5] Reis, B. F., Giné, M. F., Zagatto, E. A. G., Lima, J. L. F. C. & Lapa, R. A. (1994). Multicommutation in Flow Analysis. Part 1. Binary Sampling: Concepts, Instrumentation and Spectrophotometric Determination of Iron in Plant Digests. *Analytica Chimica Acta*, 293, 129-138.

[6] Lima, J. L. F. C., Santos, J. L. M., Dias, A. C. B., Ribeiro, M. F. T. & Zagatto, E. A. G. (2004). Multi-Pumping Flow Systems: An Automation Tool. *Talanta*, 64, 1091-1098.

[7] Miró, M., Cerdá, V. & Estela, J. M. (2002). Multisyringe Flow Injection Analysis: Characterization and Applications. *TrAC - Trends in Analytical Chemistry*, 21, 199-210.

[8] Kritsunankul, O. & Jakmunee, J. (2011). Simultaneous Determination of Some Food Additives in Soft Drinks and Other Liquid Foods by Flow Injection on-Line Dialysis Coupled to High Performance Liquid Chromatography. *Talanta*, 84, 1342-1349.

[9] Adcock, J. L., Francis, P. S., Agg, K. M., Marshall, G. D. & Barnett, N. W. (2007). A Hybrid FIA/HPLC System Incorporating Monolithic Column Chromatography. *Analytica Chimica Acta*, 600, 136-141.

[10] Sklenářová, H., Koblová, P., Chocholouš, P., Šatínský, D., Krčmová, L., Kašparová, M., Solichová, D. & Solich, P. (2011). Separation of Vitamins Retinol Acetate, Ergocalciferol, or Cholecalciferol and Tocopherol Acetate Using Sequential Injection Chromatography. *Analytical Letters*, 44, 446-456.

[11] Chocholouš, P., Šatínský, D., Sladkovský, R., Pospíšilová, M. & Solich, P. (2008). Determination of Pesticides Fenoxycarb and Permethrin by Sequential Injection Chromatography Using Miniaturized Monolithic Column. *Talanta*, 77, 566-570.

[12] Acebal, C. C., Sklenářová, H., Škrlíková, J., Šrámková, I., Andruch, V., Balogh, I. S. & Solich, P. (2012). Application of DV-SIA Manifold for Determination of Thiocyanate Ions in Human Saliva Samples. *Talanta*, 96, 107-112.

[13] Jesus, J. P., Suárez, C. A., Ferreira, J. R. & Giné, M. F. (2011). Sequential Injection Analysis Implementing Multiple Standard Additions for As Speciation by Liquid Chromatography and Atomic Fluorescence Spectrometry (SIA-HPLC-AFS). *Talanta*, 85, 1364-1368.

[14] Jiménez, M. S., Velarte, R. & Castillo, J. R. (2002). Performance of Different Preconcentration Columns Used in Sequential Injection Analysis and Inductively Coupled Plasma-Mass Spectrometry for Multielemental

Determination in Seawater. *Spectrochimica Acta Part B: Atomic Spectroscopy*, 57, 391-402.

[15] Grand, M., Oliveira, H. M., Ruzicka, J. & Measures, C. (2011). Determination of Dissolved Zinc in Seawater Using Micro-Sequential Injection Lab-on-Valve With Fluorescence Detection. *Analyst*, 136, 2747-2755.

[16] Oliveira, H. M., Segundo, M. A., Lima, J. L. F. C., Miró, M. & Cerdá, V. (2010). On-Line Renewable Solid-Phase Extraction Hyphenated to Liquid Chromatography for the Determination of UV Filters Using Bead Injection and Multisyringe-Lab-on-Valve Approach. *Journal of Chromatography A*, 1217, 3575-3582.

[17] Horstkotte, B., Duarte, C. M. & Cerdá, V. (2011). A Miniature and Field-Applicable Multipumping Flow Analyzer for Ammonium Monitoring in Seawater With Fluorescence Detection. *Talanta*, 85, 380-385.

In: Automation Systems of the 21st Century
Editors: D. Arent and M. Freebush
ISBN: 978-1-62948-262-0

Chapter 4

CROSS-COUPLING DIGITAL CONTROL OF MULTI-INDEPENDENT SERVO-DRIVES: THEORY AND EXPERIMENT

Milica B. Naumović
Faculty of Electronic Engineering, Department of Automatic Control,
University of Niš, Niš, Serbia

Abstract

The motion coordination in a system with multiple number of servomechanisms or axis has drawn much attention, especially in the modern manufacturing development. In early 1980s, the concept of cross-coupling control with a symmetrical structure was primarily introduced for the machine tool control. Quite original structures of cross-coupling speed- and position-control of the shafts of two distant electromechanical drives based on the application of the concept of the so-called electrical shaft were introduced and discussed by the author of this chapter in mid 1990s. More concretely, the proposed cross-coupling control is accomplished by the digital simulation of stiffness and friction of a virtual twisting mechanical connection between the output shafts of the drives. Since then, a number of contributions in terms of the improvement of the suggested cross-coupling concept can be found in the literature.

In this chapter, the synthesis of the structure of coordinated control in a multi drives positioning system is presented. The proposed structure is implemented in the case of three cross-coupling servomechanisms. Also, the control strategy is experimentally verified using the system for fast development and implementation of control algorithms dSPACE R&D DS1104 that provides a comfortable work in the MATLAB®/Simulink environment.

1. Introduction

Different types of cross-coupling controllers can be found in many practical applications in which the motion of two or more drives are to be coordinated in a certain sense. The synchronous operation of the drives is particularly important in paper making and processing machines, in numerically-controlled machines for metal, in nonlinear path control in automated vehicle and robot guidance, as well as in flexible systems in general.

In order to achieve a cooperation control, that since the eighties became more attractive, many different control structures have been developed. Two basic approaches can be distinguished. The first approach is the conventional distributive control where each robot is controlled separately by its own local controller, while the interactions between the robots are measured by sensors. The second approach is based on the idea of so-called cross-coupling the control systems of two distant independent servomechanisms. The first system of cross-coupling type was proposed by Sarachik and Ragazzini (1957, [1]) and had a "master-slave" structure (y follows x). Namely, the error in the y-axis affects both the x- and y- control loops, but the error signal in the x-axis is not generated (Figure 1.(a)). Such nonsymmetrical cross-coupled structure, however, requires a substantial difference in the gains for each axis, resulting in a contour error.

The concept of cross-coupling control with a symmetric structure was primarily introduced by Koren (1980, [2]) for the machine tool control. The main idea of cross-coupling control is based on the calculation of the actual contour error, multiplying it by a controller gain, and feeding the result back to the individual loops (Figure 1.(b)).

Later, many structures based on the application of cross-coupling control scheme have been proposed and tested in numerous engineering applications: biaxial control of machine tools [3], automatic guidance system of self-controlled vehicles [4], computer-controlled mobile platform for nursing or household robots [5], cross-coupling motion controller for mobile robots [6], etc.

Quite original structures of cross-coupling speed- and position-control of shafts of two distant electromechanical drives based on the application of the concept of electrical shaft are proposed in papers [7] and [8], respectively. Namely, the cross-coupling control is accomplished by the digital simulation of stiffness and friction of a virtual twisting mechanical connection between the output shafts of the drives.

The modern tools for control design, simulation, automatic code generation and hardware calibration allow an application to be developed according to the so-called "V-Cycle" (Figure 2) that is an internationally recognized development standard for IT systems. The results of a comprehensive research lasting many years at the level of modeling, design and simulation, in order to improve the structure of cross-coupling, are presented in the papers of author and contributors [9]-[17]. An effective procedure of parameter setting in digital position regulator is formulated in the paper [9]. The special observer-based structures of the system with two digitally controlled and cross-coupled servomechanisms, with induction and DC motors as actuators, are considered in papers [10]-[12]. A novel structure of the observer-based system for two manipulators which are cooperatively handling the same object in the presence of slow varying load torque disturbances is proposed in [13]. Note that two servomechanisms with cross-coupled control can be treated as a multivariable control system and then designed using different approaches [9], [14].

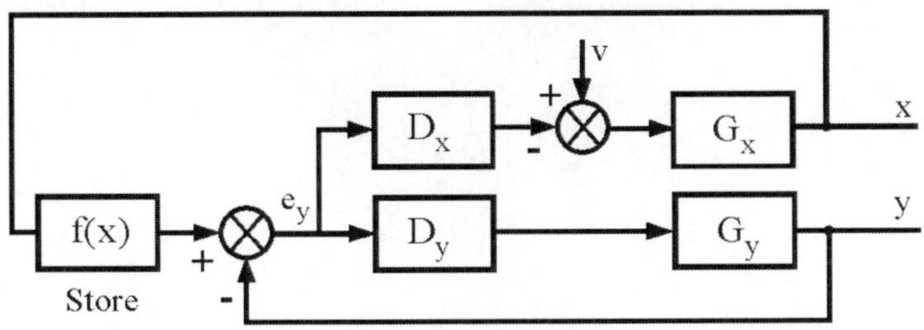

(a)

Nonsymmetrical cross-coupled system (Sarachik & Ragazzini, 1957).

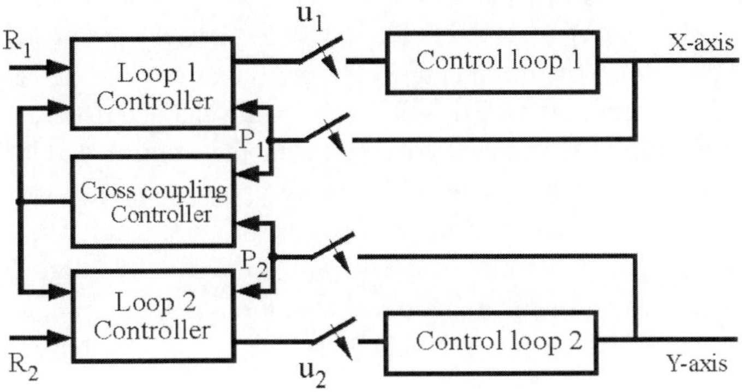

(b)

Symmetric cross-coupled control in computerized numerical control (CNC) system (Koren, 1980).

Figure 1. The block diagrams of the first controllers intended for cooperation control.

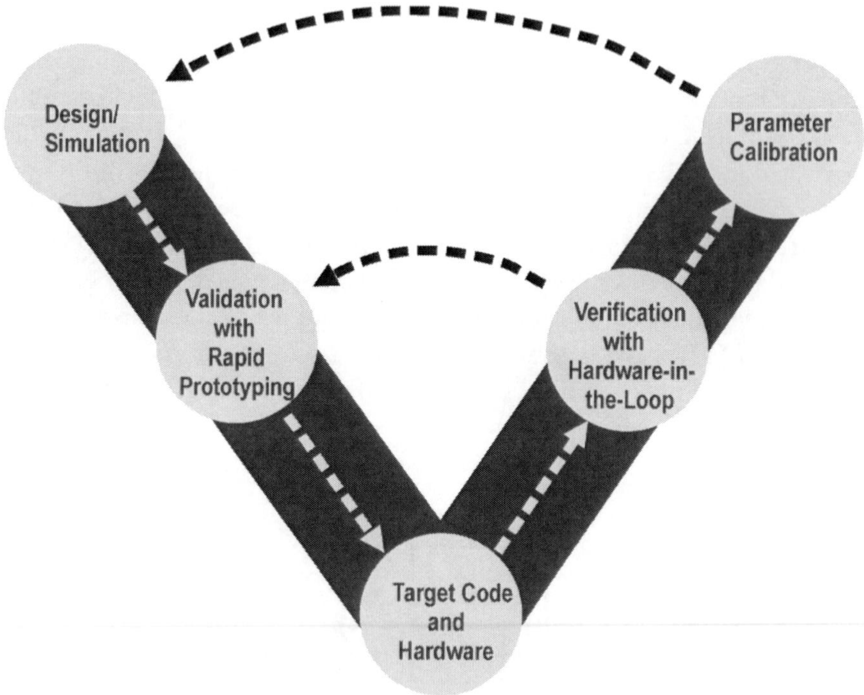

Figure 2. Implementation of design functions according to V-Circle.

The developed structure of coordinated control is implemented and experimentally verified using the system for fast development and implementation of control algorithms dSPACE R&D DS1104 that provides comfortable work in the Matlab®/Simulink environment[18].During creation of the experimental platform, in the absence of two drives with similar characteristics, one drive is modeled in Simulink environment together with the control part of the cross-coupled servo system, while a real low power permanent magnet DC motor served as the other drive. The performances of the proposed structure of the cross-coupled observer-based control system are verified by digital computer simulation, as well as experimentally in real conditions[19].

With the ever increasing demand for greater productivity, interest in synchronization of multiple motion axes or motors has been growing[20]-[23].

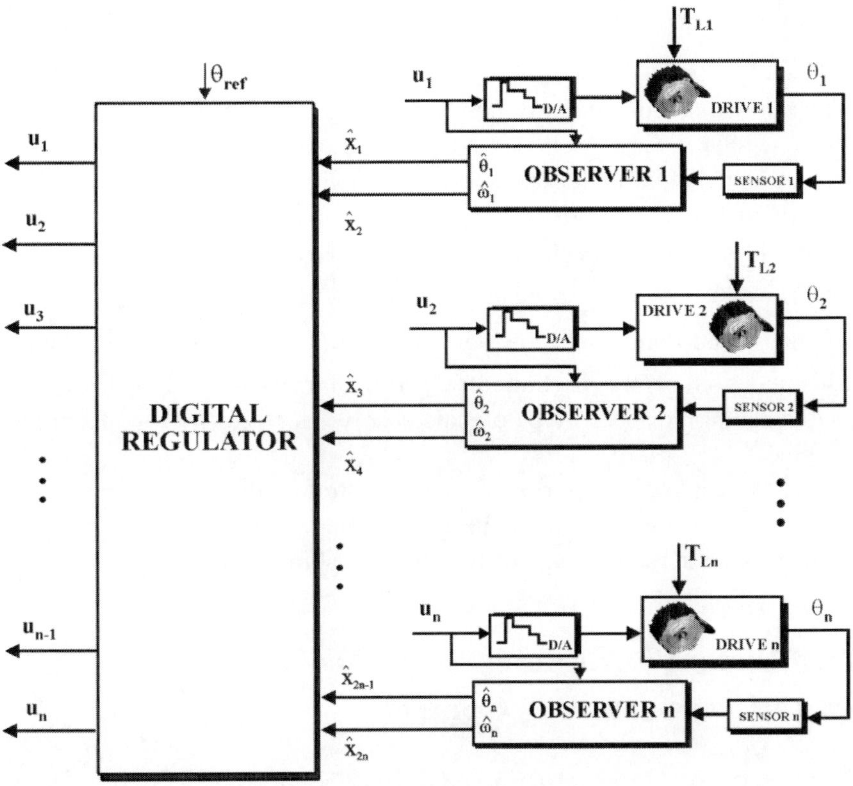

Figure 3. Blockdiagram of a servo system with cross coupling control.

2. Cross-coupling Digital Control of Multi Independent Servo-drives

The functional block diagram of the positioning servo system with the proposed cross-coupling control is shown in Figure 1. The angular positions $\theta_1(t), \theta_2(t), \ldots, \theta_n(t)$ of n distant electrical drives represent controlled variables. In the steady-state, the angular positions of drive shafts are to be the same and equal to the common reference θ_{ref}. In addition to the set point $\theta_{ref}(t) = \theta_{ref} \cdot h(t)$, the system is subjected to two kind of disturbances: load torques $T_{L1}, T_{L2}, \ldots,$ and T_{Ln} acting on drives, as well as the disturbances in the form of the initial angular displacements $\Delta\theta_i(0) = \theta_i(0) - \theta_{i+1}(0)$, $i = 1, 2, \ldots, n-1$.

The block marked in Figure 3 as "digital regulator" consists of the common position regulator, as well as of the cross-coupling mechanisms for synchronization between the drives. Note that the coupling is also established between the $n-$th and the first motor. In this way, through the signals of angular and speed differences, a special ring coupling is achieved. The angular positions $\theta_i(k)$, as well as the shaft speed signals of all n drives $\omega_i(k)$, $i = 1, 2, \ldots, n$ are adopted as coordinates of the state vector $\mathbf{x}(k)$.In order to accomplish the digital simulation of the friction of the virtual mechanical coupling between drives, the use of the special type of state observer for velocity signals estimation is discussed.

To verify the proposed ring coupled control strategy, the cross-coupling control of the positioning system with two distant servo drives is analyzed in the following section in more details with all analytical investigations.

3. Two Distant Positioning Servo Drives with Cross-coupling Control

3.1. Structure of Cross-coupling Control System

The functional block diagram of a positioning servo system with the proposed cross-coupling control [13] is shown in Figure 4. Angular positions $\theta_1(t)$ and $\theta_2(t)$ of two distant electrical drives represent controlled variables. In the steady-state, angular positions of drive shafts are to be the same and equal to the common reference θ_{ref}. The system is subjected to three kind of disturbances: set point $\theta_{ref}(t)$, initial angular displacement $\Delta\theta(0) = \theta_1(0) - \theta_2(0)$ and load torquers $T_{L1}(t)$ and $T_{L2}(t)$ acting on first and second drive, respectively.

In Figure 5, the structure of cross-coupled control system is shown in more details. The angle positions of the cross-coupled servos are controlled by the common position proportional plus integral (PI) controller. The cross-coupling mechanism comprises two elements: (i) Coupling channels that simulate the viscous friction and stiffness of a virtual mechanical shaft coupling the servos; (ii) Digital PIcontroller that eliminates the difference between the steady-state values of angular positions of the servos. Such kind of cross-coupling control based on the digital simulation of torques of a virtual mechanical link between the shafts of servomechanisms, represents the main feature of the suggested structure.

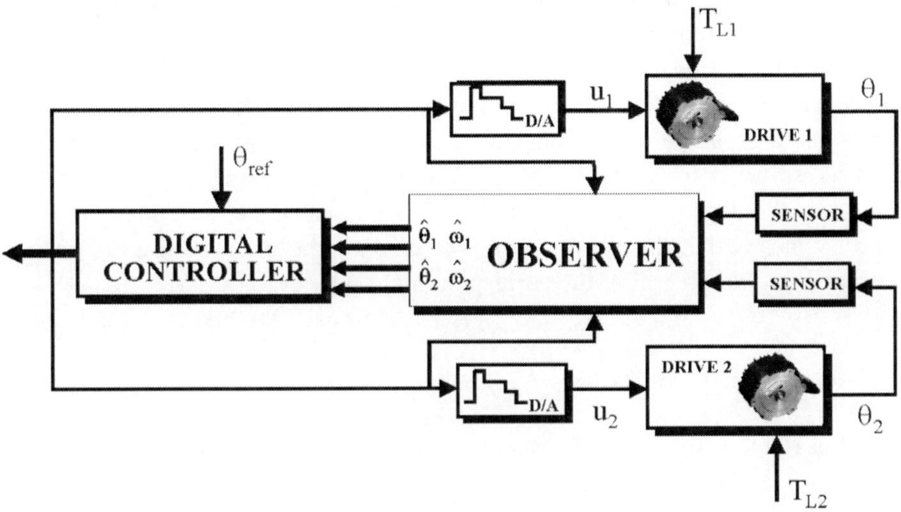

Figure 4. Blockdiagram of a positioning servo system with cross coupling control.

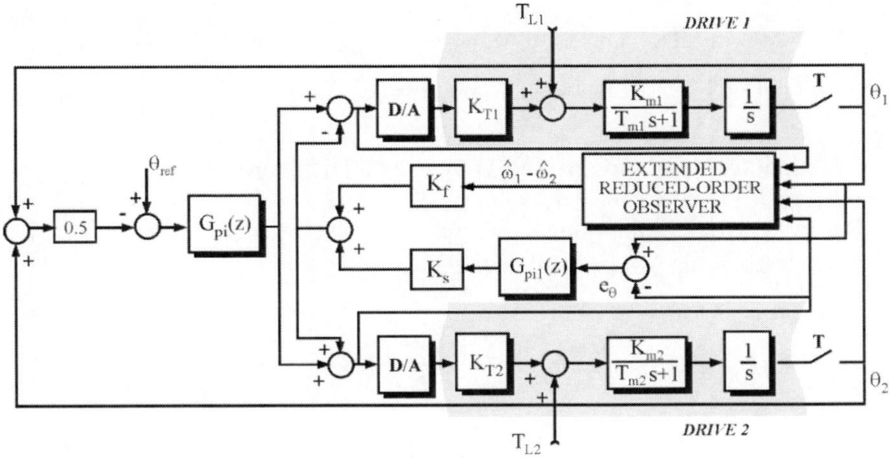

Figure 5. Structure of cross-coupling control system.

In the considered system six control parameters are adjustable: gains in the transfer function of digital PI-type controller $G_{pi}(z)$, the friction and stiffness parameters, and gains in the transfer function $G_{pi1}(z)$ of the additional PI-type controller in the local feedback loop. The control objectives are set to obtain the desired speed of the system response and stability margin.

The use of the observer within the scheme of the cross-coupling control system is illustrated in Figure4. The observer, as a special dynamic system, processes the torque commands and the quantized signals from the position transducers and enables the correct estimation of shaft position and speed signals in the presence of the constant or slow varying load torque disturbances.

It is suitable to discuss the possibility of employing the reduced-order observer with additional state integrators [26]. The attribute "extended" in Figure 5 exactly means that the applied observer is obtained by extending the ordinary observer with the integral of the corresponding state variables.

3.2. Steady-state Analysis

For a stable cross-coupled control system, it is of interest to investigate the steady-state error $e_\theta(\infty)=\theta_1(\infty)-\theta_2(\infty)$ in dependence on the set-point $\theta_{\text{ref}}(t)=\theta_{\text{ref}}\,\text{h}(t)$ and load torques $T_{L1}(t)$ and $T_{L2}(t)$. For the sake of simplicity in further developments, the signals $T_{L1}^*(t)$ and $T_{L2}^*(t)$ instead of $T_{L1}(t)$ and $T_{L2}(t)$ are introduced; that is $T_{L1}^*(t)=T_{L1}(kT)$ and $T_{L2}^*(t)=T_{L2}(kT)$ for $kT\le t<(k+1)T$, $k=0,1,2,\ldots$

3.2.1. Characteristics of Coupled Drives Are Different

The system block-diagram in Figure 5 may be considered as a signal-flow graph. Then, applying Mason's rule, we get

$$
\begin{aligned}
E_\theta(z)=G_{pi}(z)&\frac{G_3(z)\big(1+2K_fG_2(z)\big)-G_4(z)\big(1+2K_fG_1(z)\big)}{F(z)}\Theta_{ref}(z)\\
&+\frac{G_3(z)\big(1+G_{pi}(z)G_4(z)+K_fG_2(z)\big)-K_fG_1(z)G_4(z)}{F(z)}\frac{T_{L1}^*(z)}{K_{T1}}\\
&-\frac{G_4(z)\big(1+G_{pi}(z)G_3(z)+K_fG_1(z)\big)-K_fG_2(z)G_3(z)}{F(z)}\frac{T_{L2}^*(z)}{K_{T2}}\,,
\end{aligned}
\tag{1}
$$

...tting the Controller Parameters

...adjustable control parameters ($K_p, K_I, K_f, K_s, K_{p1}$ and K_{I1}) of the ...iven in Figure 5 can be set into three steps.

...ettingCommon PI Regulator Parameters

...parameters of the position PI-type regulator (K_p, K_I) are to be set in ...ce with the desired quality of system continuous-time responses to the ...ommon reference signal. ..., consider the characteristic equation of the decoupled system. Thus, ...$_f = K_s = 0$ into (2) yields

$$F(z) = 1 + 0.5G_{pi}(z)\left(G_3(z) + G_4(z)\right), \tag{11}$$

...nsfer functions $G_{pi}(z)$, $G_3(z)$ and $G_4(z)$ are given by Equations (3), ...5). After simple manipulations, (11) can be reduced into fourth-order ...l form. In applying the pole placement method, the speed of ...s-time response and stability margin are specified by the natural ...ω_n and the damping ratio ζ of the dominant poles ...$\omega_n \pm j\omega_n\sqrt{1-\zeta^2}$ located inside the principal strip of the s-plane. Let ...that the other two poles are roots of equation

$$z^2 + A_1^* z + A_0^* = 0. \tag{12}$$

...id their great influence on the continuous-time response of the closed-..., the poles must be placed inside the unit circle and very close to the ...z-plane, and thus holds

$$\left|A_0^*\right| < 1, 1 + A_1^* + A_0^* > 0, \text{ and } 1 - A_1^* + A_0^* > 0. \tag{13}$$

where $\Theta_{ref}(z)$ denotes z-transform of reference signal; $T_{L1}^*(z)$ and $T_{L2}^*(z)$ are z-transforms of load torque disturbances acting on first and second drive, respectively. In above equation $F(z)$ represents the determinant of the graph, given as

$$\begin{aligned} F(z) = 1 + 0.5G_{pi}(z)\left(G_3(z)+G_4(z)\right) + K_f\left(G_1(z)+G_2(z)\right) + K_k\left(G_3(z)+G_4(z)\right) \\ + G_{pi}(z)\left(2K_kG_3(z)G_4(z) + K_fG_1(z)G_4(z) + K_fG_2(z)G_3(z)\right) \end{aligned} . \tag{2}$$

The transfer functions appearing in Equations (1) and (2) are calculated as follows:

$$G_{pi}(z) = K_p + \frac{K_I z}{z-1}; \tag{3}$$

$$G_1(z) = \mathrm{Z}\left\{\frac{1-e^{-sT}}{s}\frac{K_{T1}K_{m1}}{T_{m1}s+1}\right\} = \frac{K_1(1-A)}{z-A}$$

and

$$G_2(z) = \mathrm{Z}\left\{\frac{1-e^{-sT}}{s}\frac{K_{T2}K_{m2}}{T_{m2}s+1}\right\} = \frac{K_2(1-B)}{z-B}, \tag{4}$$

$$A = \exp(-T/T_{m1}),$$

$$B = \exp(-T/T_{m2}),$$

$$K_1 = K_{T1}K_{m1},$$

$$K_2 = K_{T2}K_{m2};$$

$$G_3(z) = \mathrm{Z}\left\{\frac{1-e^{-sT}}{s}\frac{K_{T1}K_{m1}}{s(T_{m1}s+1)}\right\} = \frac{b_{01}z + b_{11}}{(z-1)(z-A)} \tag{5}$$

$$b_{01}=K_1\left[T-(1-A)T_{m1}\right],\ b_{11}=K_1\left[-TA+(1-A)T_{m1}\right];$$

and

$$G_4(z)=\mathcal{Z}\left\{\frac{1-e^{-sT}}{s}\frac{K_{T2}K_{m2}}{s\left(T_{m2}s+1\right)}\right\}=\frac{b_{02}z+b_{12}}{(z-1)(z-B)} \tag{6}$$

$$b_{02}=K_2\left[T-(1-B)T_{m2}\right],\ b_{12}=K_2\left[-TB+(1-B)T_{m2}\right]$$

In above equations the parameters K_{Ti}, K_{mi} and T_{mi}, $i=1,2$ are the constants of the electrical driving torques, as well as gain factors and mechanical time constants of the considered drives, respectively. The zero-order hold-equivalence method is known in the literature as a suitable method for the discrete equivalent derivation of a continuous-time system transfer function, particularly in the case of digital hardware implementation [24].

On the basis of mentioned relationships, one can conclude that the function $F(z)$, given by (2), has the third-order pole at the point $z=1$. The same inspection leads to the conclusion that numerators of second and third terms on the right-hand side of Equation (1) have triple poles at $z=1$, too. Consequently, in the presence of the constant step reference signal $\theta_{ref}(t)=\theta_{ref}\,h(t)$ and constant load torques $T^*_{L1}(t)=T_{L10}\,h(t-\tau_1)$ and $T^*_{L2}(t)=T_{L20}\,h(t-\tau_2)$, the steady-state error is calculated using the final-value theorem as

$$e_\theta(\infty)=\frac{1}{2K_s}\left(\frac{T_{L10}}{K_{T1}}-\frac{T_{L20}}{K_{T2}}\right). \tag{7}$$

Hence, in the steady-state, angle positions of the cross-coupled servos are equal to each other and to the common set-point only in the absence of disturbances. However, in the present of load torques having the different values, the steady-state position error (7) exists. Thus, a virtual mechanical shaft which couples the drives is twisted by the angle $e_\theta(\infty)$ to compensate the difference between load torques.

3.2.2. Characteristics of Coupled Drives Are Identical

In the case of identical drives, it is convenient to define

$$G_1(z)=G_2(z)=G(z),$$

and

$$G_3(z)=G_4(z)=G^\circ(z).$$

The characteristic polynomial $F(z)$, given by (2), ca

$$F^\circ(z)=\left(1+G_{pi}(z)G^\circ(z)\right)\left(1+2K_fG(z)\right.$$

We note that the pole spectrum of the closed-lo
Namely, it consists of two pole pairs that can be placed s
From (1) and (8) it follows that the z-transfc
independent on reference signal so that

$$E^\circ_\theta(z)=\frac{G^\circ(z)\left(1+G_{pi}(z)G^\circ(z)\right)}{F^\circ(z)K_T}\left(T^*_{L1}\right. [illegible]$$

$$K_T=K_{T1}=K_{T2}.$$

Hence, in the steady-state, the angle positions of
other regardless of constant load torques having the sa

3.2.3. Steady-state Position Error Compensatio

In the classical control system design, it is well
used to eliminate steady-state errors in the present
external disturbances. Thus, to eliminate error (7
introduced within the structure of local feedback loc
the stiffness of the virtual mechanical coupling (see

Using the outlined pole placement procedure, the control parameters K_p and K_I can be calculated from

$$\begin{bmatrix} 0 & d_1 & 0 & -1 \\ -d_1 & -d_2 & -1 & z_1+z_2 \\ d_2 & -d_3 & z_1+z_2 & -z_1 z_2 \\ d_3 & 0 & -z_1 z_2 & 0 \end{bmatrix} \begin{bmatrix} P_1 \\ P_2 \\ A_0^* \\ A_1^* \end{bmatrix} = \begin{bmatrix} -(z_1+z_2)+2+A+B \\ z_1 z_2 - 1 - 2A - 2B - AB \\ A+B+2AB \\ -AB \end{bmatrix}, \quad (14)$$

where

$$P_1 = K_p, P_2 = K_p + K_I \quad (15)$$

and

$$d_1 = 0.5(b_{01} + b_{02})$$

$$d_2 = 0.5\left[(Bb_{01} - b_{11}) + (Ab_{02} - b_{12})\right] \quad (16)$$

$$d_3 = 0.5(Bb_{11} + Ab_{12}).$$

3.3.2. Choice of Cross-coupling Parameters

It is shown that if the drives of the control system in Figure 5 are identical, setting the controller parameters is based on the special separation principle. In virtue of (9), the common PI-type regulator parameters and cross-coupling control parameters can be adjusted independently. Namely, only two roots of (9) are determined by cross-coupling parameters K_f and K_s. To avoid significant changes in the closed-loop system continuous-time response, the cross-coupling dynamics must be adjusted to be two to four times faster then the dynamics of the closed-loop control system. In majority of cross-coupling control system, characteristics of coupled drives are not identical but very close to each other. In such a case, appropriate poles are weakly coupled, and the procedure of adjusting control parameters can be carried out in the same way as in the previous subsection.

For control parameter setting, the worst but seldom case arises when coupled drives are quite different. Then, using the outlined pole placement procedure parameters K_p and K_I are first calculated. After setting numerical values of fix parameters of the considered drives as well as calculated values of parameters K_p and K_I into (2), one obtains the system characteristic equation with coefficients that depend only on K_f and K_s. Inspecting pairs $\left(K_f, K_s\right)$ for which obtained characteristic equation has all roots inside the unit circle of z-plane, we determine the stable region of $\left(K_s, K_f\right)$-plane. After several simulation runs, the appropriate values of cross-coupling parameters belonging to the determined stable region of $\left(K_s, K_f\right)$-plane can be selected.

3.3.3. Setting Parameters of the Additional PI-Type Regulator

To relax torsion tension of the virtual mechanical coupling, a local feedback loop of the considered system was modified by adding a PI-type regulator (Figure 5). Setting the parameters of this additional regulator can be carried out in the last step. To retain the stiffness factor value adjusted in previous step, the proportional gain is adopted to be $K_{p1} = 1$. It is convenient to choose the value of the gain of the integral action K_{I1} as the ratio of determined parameters K_I and K_p.

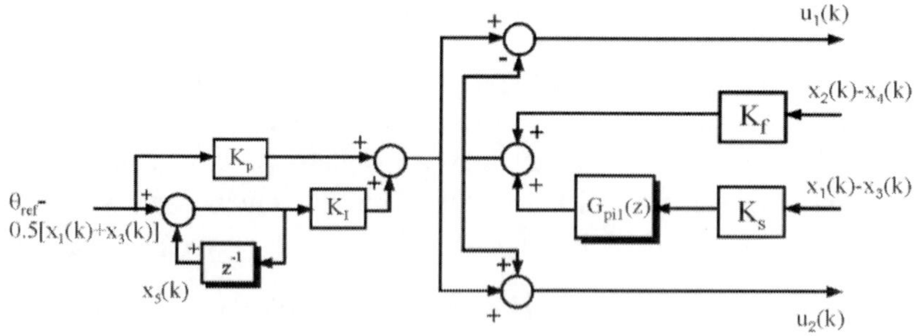

Figure 6. Simulation diagram of digital regulator given in Figure 4.

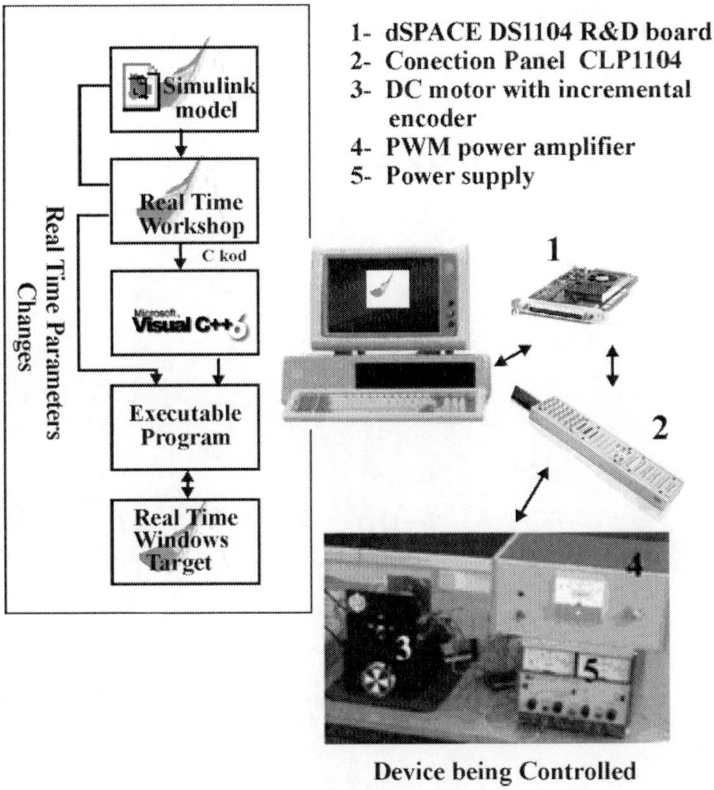

Figure 7. Schematic representation of experimental setup.

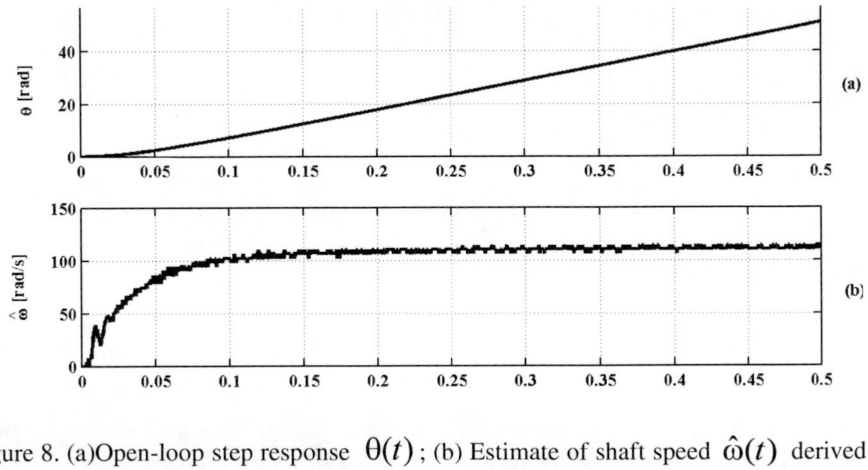

Figure 8. (a)Open-loop step response $\theta(t)$; (b) Estimate of shaft speed $\hat{\omega}(t)$ derived by Euler's approximation of the derivative.

3.4. The State-space Approach to Cross-coupling Control Design

The state-space description of the decoupled system $\left(K_f = K_s = 0\right)$ is

$$\mathbf{x}(k+1) = \mathbf{E}\mathbf{x}(k) + \mathbf{f}u(k) + \mathbf{I}\theta_{ref} \tag{17}$$

$$\mathbf{c}(k) = \mathbf{d}\mathbf{x}(k), \tag{18}$$

where

$$\mathbf{E} = \left[\begin{array}{cc:cc:c} \multicolumn{2}{c:}{\mathbf{E}_1(T)} & \multicolumn{2}{c:}{\mathbf{0}_{2\times2}} & \mathbf{0}_{2\times1} \\ \hdashline \multicolumn{2}{c:}{\mathbf{0}_{2\times2}} & \multicolumn{2}{c:}{\mathbf{E}_2(T)} & \mathbf{0}_{2\times1} \\ \hdashline -0.5 & 0 & -0.5 & 0 & 1 \end{array}\right], \quad \mathbf{f} = \left[\begin{array}{c} \mathbf{f}_1(T) \\ \hdashline \mathbf{f}_2(T) \\ \hdashline 0 \end{array}\right], \tag{19}$$

$$\mathbf{d} = \begin{bmatrix} 1 & 0 & 0 & 0 & 0 \\ 0 & 0 & 1 & 0 & 0 \end{bmatrix}, \mathbf{I} = \begin{bmatrix} 0 & 0 & 0 & 0 & 1 \end{bmatrix}^{\mathrm{T}}, \tag{20}$$

and

$$\mathbf{x}(k) = \left[\mathbf{x}_1(k) \;\vdots\; \mathbf{x}_2(k) \;\vdots\; x_5(k)\right]^{\mathrm{T}} = \left[\theta_1(k) \quad \omega_1(k) \;\vdots\; \theta_2(k) \quad \omega_2(k) \;\vdots\; x_5(k)\right]^{\mathrm{T}} \tag{21}$$

The related matrices $\mathbf{E}_i(T)$, $\mathbf{f}_i(T)$, $i = 1, 2$ are given as follows:

$$\mathbf{E}_i(T) = \begin{bmatrix} 1 & T_{mi}\left(1 - \mathrm{e}^{-T/T_{mi}}\right) \\ 0 & \mathrm{e}^{-T/T_{mi}} \end{bmatrix}, \mathbf{f}_i(T) = \begin{bmatrix} K_{Ti}K_{mi}\left(T + T_{mi}\,\mathrm{e}^{-T/T_{mi}} - T_{mi}\right) \\ K_{Ti}K_{mi}\left(1 - \mathrm{e}^{-T/T_{mi}}\right) \end{bmatrix}. \tag{22}$$

In above equations the parameters K_{Ti}, K_{mi} and T_{mi}, $i = 1, 2$ have the same meaning as in (4)-(6).

Suppose that all state variables are measurable, i.e. it is possible to form a total negative feedback as shown in Figure 6.

$$u(k) = -\mathbf{K}_1{}^{\mathrm{T}}\mathbf{x}(k) + u_{ref}, \tag{23}$$

where

$$\mathbf{K}_1{}^{\mathrm{T}} = \left[0.5\left(K_p + K_I\right) \;\vdots\; 0 \;\vdots\; 0.5\left(K_p + K_I\right) \;\vdots\; 0 \;\vdots\; -K_I \right] \tag{24}$$

and

$$u_{ref} = \left(K_p + K_I\right)\theta_{ref}. \tag{25}$$

The characteristic polynomial of the decoupled closed-loop system is given by

$$\Delta\left(z, \mathbf{K}_1\right) = \det\left(z\mathbf{I} - \mathbf{E} + \mathbf{f}\mathbf{K}_1{}^{\mathrm{T}}\right). \tag{26}$$

The state-space difference equation of the cross-coupled system is derived as

$$\mathbf{x}(k+1) = \mathbf{E}\mathbf{x}(k) + \mathbf{F}\mathbf{u}(k) + \mathbf{I}\theta_{ref} \tag{27}$$

with

$$\mathbf{F} = \left[\begin{array}{c:c} \mathbf{f}_1(T) & \begin{matrix} 0 \\ 0 \end{matrix} \\ \hdashline \begin{matrix} 0 \\ 0 \end{matrix} & \mathbf{f}_2(T) \\ \hdashline 0 & 0 \end{array} \right],$$

and

$$\mathbf{u}(k) = \left[\begin{array}{c} u_1(k) \\ \hdashline u_2(k) \end{array} \right]. \tag{28}$$

In such a system with full state feedback control

$$\mathbf{u}(k) = -\mathbf{K}_2{}^{\mathrm{T}}\mathbf{x}(k) + \begin{bmatrix} u_{ref} \\ u_{ref} \end{bmatrix} \tag{29}$$

the gain matrix and the corresponding characteristic polynomial are given respectively as follows:

$$\mathbf{K}_2{}^{\mathrm{T}} = \left[\begin{array}{c|c|c|c|c} 0.5\left(K_p + K_I\right) + K_k & K_f & 0.5\left(K_p + K_I\right) - K_k & -K_f & -K_I \\ \hline 0.5\left(K_p + K_I\right) - K_k & -K_f & 0.5\left(K_p + K_I\right) + K_k & K_f & -K_I \end{array}\right] \tag{30}$$

and

$$\Delta\left(z, \mathbf{K}_2\right) = \det\left(z\mathbf{I} - \mathbf{E} + \mathbf{F}\mathbf{K}_2{}^{\mathrm{T}}\right). \tag{31}$$

Recall, in the case of identical drives, the pole spectrum of the considered cross-coupled system is decoupled; it consists of two pole pairs that can be placed separately. Namely, common PI-type regulator parameters (K_p, K_I) and cross-coupling control parameters (K_f, K_s) can be adjusted independently. In the case of drives with identical characteristics, it is convenient to define

$$\mathbf{E}^{\circ} = \mathbf{E}_i(T)\ ,\ \ \mathbf{f}^{\circ} = \mathbf{f}_i(T),\ \ i = 1, 2 \ . \tag{32}$$

In that case, the characteristic polynomial can be rewritten as

$$\Delta^{\circ}\left(z, \mathbf{K}_2\right) = \Delta_{\mathrm{K_s K_f}}(z) \cdot \Delta_{\mathrm{PI}}(z) \tag{33}$$

where

$$\Delta_{\mathrm{K_s K_f}}(z) = 1 + \left[\begin{array}{c|c} 2K_s & 2K_f \end{array}\right]\left(z\mathbf{I} - \mathbf{E}^{\circ}\right)^{-1}\mathbf{f}^{\circ} \tag{34}$$

and

$$\Delta_{\mathrm{PI}}(z) = 1 + \left[\begin{array}{c|c} \left(K_p + K_I\right) + K_I (z-1)^{-1} & 0 \end{array}\right]\left(z\mathbf{I} - \mathbf{E}^{\circ}\right)^{-1}\mathbf{f}^{\circ} \tag{35}$$

In majority of cross-coupling control system, characteristics of coupled drives are not identical but very close to each other. In such a case, the mentioned pole pairs are weakly coupled.

3.5. Position and Velocity Observation

The observer associated with the drives in Figure 4 estimates the states $\mathbf{x}_1(k)$ and $\mathbf{x}_2(k)$ in (21) by using the control variables and measured outputs of the considered system. Note that angular positions in system in Figure 4 are measurable, but it is suitable to apply PI^2 observers [26], bearing in mind the effects of full observer implementation from the viewpoint of filtering of measurement noise, as well as in order to enable the correct estimation of the state variables in the presence of constant or slow varying load torque disturbances.

After adding new state variables $\hat{x}_3(k)$ and $\hat{x}_4(k)$ into state vector to be observed $\hat{\mathbf{x}}(k)=\begin{bmatrix}\hat{\theta}(k) & \hat{\omega}(k)\end{bmatrix}^{\mathrm{T}}$, the vector difference equation of the suggested observer becomes

$$\hat{\mathbf{x}}_{\mathrm{e}}(k+1)=\left(\mathbf{E}_{\mathrm{e}}-\mathbf{G}_{\mathrm{e}1}\mathbf{D}_{\mathrm{e}}\right)\hat{\mathbf{x}}_{\mathrm{e}}(k)+\mathbf{f}_{\mathrm{e}}u(k)+\mathbf{G}_{\mathrm{e}2}\begin{bmatrix}c(k)\\ c(k-1)\end{bmatrix}, \tag{36}$$

where

$$\hat{\mathbf{x}}_{\mathrm{e}}(k+1)=\begin{bmatrix}\hat{\mathbf{x}}(k+1)\\ \hat{x}_3(k+1)\\ \hat{x}_4(k+1)\end{bmatrix},\quad \mathbf{E}_{\mathrm{e}}=\begin{bmatrix}1 & e_1 & 1 & 0\\ 0 & e_2 & 0 & 1\\ 0 & 0 & 1 & 0\\ 0 & 0 & 0 & 1\end{bmatrix},$$

$$\mathbf{f}_{\mathrm{e}}=\begin{bmatrix}f_1\\ f_2\\ 0\\ 0\end{bmatrix},\quad \mathbf{D}_{\mathrm{e}}=\begin{bmatrix}1 & 0 & 0 & 0\\ 0 & 1 & 0 & 0\end{bmatrix}, \tag{37}$$

$$\mathbf{G}_{\mathrm{e}1}=\begin{bmatrix}g_1 & 0\\ g_2 & 0\\ g_3 & 0\\ 0 & g_4\end{bmatrix},\quad \text{and}\quad \mathbf{G}_{\mathrm{e}2}=\mathbf{G}_{\mathrm{e}1}\begin{bmatrix}0 & 0\\ 1/T & -1/T\end{bmatrix} \tag{38}$$

The observer gain matrix $\mathbf{G}_{e1}$ in (36) is calculated according to the desired observer pole spectrum determining the speed of convergence between the state of the system and the state estimated by the observer. The setting of gains g_i , $i = 1, 2, 3, 4$ may be conveniently performed assuming all observer poles to be the same and equal to $\sigma_z = \exp(-2\pi f_0 T)$, where f_0 denotes the observer bandwidth [25]. It is shown in [26] with more details that the observer gains are calculated from

$$\det\begin{bmatrix} z-1+g_1 & -e_1 & -1 & 0 \\ g_2 & z-e_2 & 0 & -1 \\ g_3 & 0 & z-1 & 0 \\ 0 & g_4 & 0 & z-1 \end{bmatrix} = \left(z-\sigma_z\right)^4 . \tag{39}$$

4. Experimental Setup for Verification of Cross-Coupling Control Algorithm

To illustrate and verify the effectiveness of the proposed cross-coupling control strategy discussed in previous section, the example of the system with three cross-coupling servomechanisms with quite different characteristics is considered. Figure 7 visualizes the principal scheme of the experimental environment for rapid control prototyping.

The real system in Figure 7 is composed of a DC motor with incremental encoder, and PWM power amplifier with power supply. Note that the real control plant is the low power DC motor with some dry friction problems, which are especially expressive in the tasks of positioning. The other two drives have been implemented in the Matlab®/Simulink environment, and in this way the change of the parameter values in their transfer functions is enabled. Also, the control part of the proposed cross-coupled system and state observers are implemented in the same environment.

Notice that the coordination in the considered servo system, whose experimental verification is possible in the environment shown in Figure 7, is based on the assumption that the motion of all motors is coordinated if every pair of motors is coordinated in their motion. Recall that the considered system with cross coupling control has a number of adjustable parameters – six control parameters and certain number of gain values of the applied full-order or reduced-order observers. It has been shown that in the case of identical drives, the cross coupling control parameters (K_f , K_s) can be adjusted independently from the position regulator parameters. The procedure for setting parameters of the digital regulator is simple and can be carried out in three steps, while the observer gains are determined in the fourth step.

For the sampling period the value $T = 0.001\,\text{s}$ is adopted. The speed of continuous-time closed-loop system responses and stability margin are specified by the dominant pole pair (the relative damping coefficient is $\zeta = 0.707$ and the natural frequency is $\omega_n = 10$ rad/s) located in Nyquist frequency region[27].

On the basis of the experimentally recorded step response of the servo drive with low power permanent magnet DC motor given in Figure8, the parameters in transfer function (4) can be determined as $K_1 = 22.065$, and $T_{m1} = 0.0348\,\text{s}$ [28].

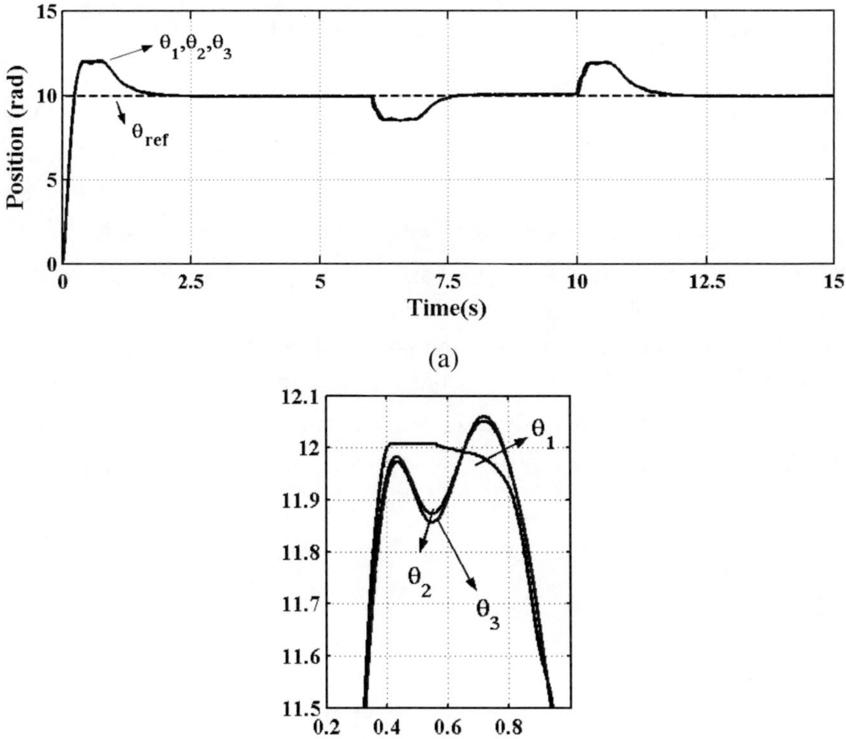

Figure 9. (a)Responses of angular positions to both step reference signal and disturbance signal;(b) Magnified part of the responses given in Figure 9(a).

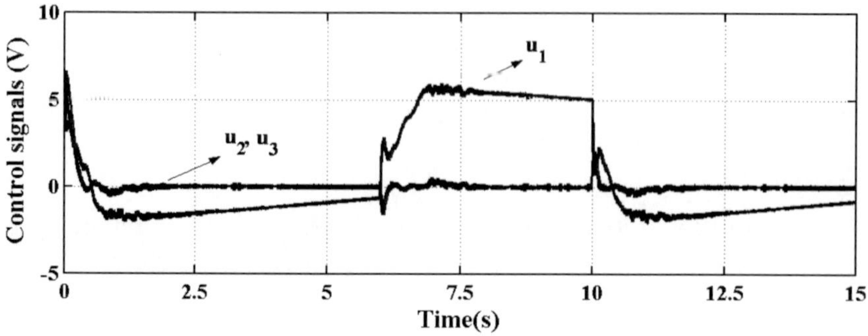

Figure 10. Control signals.

The system is excited by the common step reference signal $\theta_{\text{ref}} = 10 \cdot \text{h}(t)\,\text{rad}$. A constant load torque disturbance signal $T_{\text{L}1} = -0.1 \cdot \{\text{h}(t-6) - \text{h}(t-10)\}\,\text{Nm}$, acting on the first drive in a certain period of time, amounts to 53% of the nominal torque value of the applied motor.

The traces of the angular positions of the drives in Figure 9 show that the response quality is preserved. By using the cross-coupling strategy, the desired effect is achieved, i.e. the steady-state values of the angular positions of all three drive shafts are the same and equal to the common reference signal θ_{ref}. The achieved quality of continuous-time angular responses and speed of state estimation by the applied observers are in agreement with results expected in analytical design of the system. Also, the ability of observers to estimate both the shaft positions and speeds, when the constant load torque exists, is visualized in Figure 9. Note that the control signals u_1, u_2 and u_3 in Figure 10 are without expressive chattering and in agreement with the results of the coordinated motion of the drives.

Conclusion

Quite original structures of cross coupling speed and position control of the shafts of two distant electromechanical drives based on the application of the concept of the electrical shaft are proposed by Stojić and Naumović in mid 1990s [12]. The structure enables synchronous motion of the drives by velocity and angle positions of their shafts. The suggested cross-coupling control is based upon the idea of simulation of effects that appear in a virtual mechanical link between the shafts of servomechanisms. System steady-states are analyzed in the presence of constant reference signal and load torque disturbances. The design procedures of common position controller, cross-coupling control, and observer are proposed. Since the certain portions of controlling mechanism are decoupled or weakly coupled, the control parameters may be tuned by using a relatively simple procedure that can be applied in both the similar and quite different servo-drives. The results of some experimental runs in the environment for rapid control prototyping are presented in this chapter, to show the validity of the proposed control strategy. The efficiency of the special structure of multiple positioning servomechanisms with cross-coupling control is experimentally verified in the case of three cross-coupling servomechanisms.

Acknowledgments

Research in this area is supported in part by the Project Grant III44004 (2011-2014) financed by Ministry of Education, Science and Technological Development, Republic of Serbia.

References

[1] Sarachik P., Ragazzini J.R., A Two Dimensional Feedback Control System, *Trans. AIEE* 76, 1957, pp. 55-61.

[2] Koren Y., Cross-coupled biaxial computer control for manufacturing systems, *Journal of the Dynamic Systems, Measurement and Control,* vol. 102, 1980, pp. 265-272.

[3] Srinivasan K., Kulkarni P.K., Cross-coupled control of biaxial feed drive servomechanism, *ASME J. Dynamic Syst., Meas., Control,* vol. 112, no. 2, June 1990, pp.225-232.

[4] Hongo T., Arakawa H., Sugimoto G., Tange K., Yamamoto Y., An autonomous guidance system of a self-controlled vehicle, *IEEE Trans. Ind. Electron* 34 (1), 1987, pp. 1772-1778.

[5] Borestein J., Koren Y., A mobile platform for nursing robots, *IEEE Trans. Ind. Electron 32* (2), 1985, pp. 158-165.

[6] Feng L., Koren Y., Borenstein J., Cross-coupling motion controller for mobile robots, *IEEE Control Systems,*December 1993, pp.35-43.

[7] Stojić M. R., Vukosavić S. N., and Stojić Đ. M., Design of digitally drives coupled by an electrical shaft. In *Proc. of the Symposium on Power*

Electronics Ee'95, (Novi Sad, Yu), 1995, pp. 55-63 (invited paper in Serbian).

[8] Stojić M. R., Vukosavić S. N., and Naumović M. B., Positioning servodrives with cross-coupling control, Part I: Structure design. In *Proc. XL ETRAN Conference,* (Budva, Yu), Society for ETRAN, Beograd, 1996, pp. 533-536(in Serbian).

[9] Naumović M. B. and Stojić M. R., Positioning servodrives with cross-coupling control, Part II: Setting of parameters. In *Proc. XL ETRAN Conference,* (Budva, Yu), Society for ETRAN, Beograd, 1996, pp. 537-540(in Serbian).

[10] Stojić M. R., Naumović M. B., and Vukosavić S. N., Cross-coupled speed-controlled drives with Tesla's induction motor. In *Proc. of the 5th Inter Conference TESLA III MILLENNIUM,* (Beograd, Yu), SASA, Beograd, 1996, pp. II.51-II.64 (Invited paper).

[11] Naumović M. B. and Stojić M. R., Positioning servodrives with cross-coupling control, Part III: Multivariable Approach in System Analysis and Design, In *Proc. XLI ETRAN Conference,* (Zlatibor, Yu), Society for ETRAN, Beograd, 1997, pp. 401-404 (in Serbian).

[12] Naumović, M. B. and Stojić M. R., Design of the Observer-Based Cross-Coupled Positioning Servo-drives, In *Proc. of the IEEE International Symposium on Industrial Electronics ISIE'97,* (Guimarães, Portugal), IEEE, Piscataway, N.J., 1997, pp. 643-648.

[13] Naumović M. B., Cross-Coupled Motion Controller for Two Cooperating Robot Arms. In *Proc. of the IEEE International Symposium on Industrial Electronics ISIE'99,* (Bled, Slovenia), IEEE, Piscataway, N.J., 1999, pp. 909-913.

[14] Naumović M. B. and Stojić M. R., Two Distant Positioning Servodrives with Cross-Coupling Control, Transactions on Automatic Control and Computer Science-Special Issue dedicated to 6th International Conference on Technical Informatics, CONTI2004, *Timisoara,* May 27-28, 2004, vol.49(63), No. 1, pp. 35-40.

[15] Naumović M. B., Tracking in digital systems with cross-coupling control, In *Proc. XL ETRAN Conference,* (Čačak, SCG), Society for ETRAN, Beograd, 2004, pp. 215-218(in Serbian).

[16] Naumović M. B. and Stojić M. R., Tracking in Digital Systems with Cross-Coupling Control, Preprints of *the 4th International IFAC Workshop on Automatic Systems for Bilding the Infrastructure in Developing Countries, DECOM'04*, Bansko, Bulgaria, 2004, pp. 173-178.

[17] Naumović M. B. and Stojić M. R., Regulation and tracking in digital systems with cross-coupling control, In *Proc. INFOTEH-JAHORINA Symposium,*2005, Vol. 4, Ref. A-13, pp. 58-62 (in Serbian).

[18] *dSPACE International*. Available: http://www.dspaceinc.com/en/inc/home.cfm.

[19] Naumović M. B., and Stojić M. R., *Two Distant Cross-coupled* Positioning *Servo Drives*: Theory And Experiment, Electronics, University of Banja Luka, Vol. 13, No. 2, December 2009, pp. 25-29.

[20] Sun D., Position synchronization control of multiple motion axes with adaptive coupling control, *Automatica*, 39, 2003, pp. 997–1005.

[21] Xiao Y., Zhu K., Liaw H. C., Generalized synchronization control of multi-axis motion systems, *Control Engineering practice*, 13, 2005, pp. 809–819.

[22] Xiao Y., Zhu K., Optimal Synchronization Control of High-Precision Motion Systems, *IEEE Trans. Ind. Electron* 53 (4), 2006, pp. 1160-1169.

[23] Cheng M. H., Chen Ch-Yi, Bakhoum E. G., Synchronization Controller Synthesis of Multi-axis Motion System, *International Journal of Innovative Computing, Information and Control*, Vol. 7, No. 7(B), 2011, pp. 4395-4410.

[24] Naumović M. B., A guide to model parameter conversion in a dual-rate adaptive control system, Proceedings of the Institution of Mechanical Engineers, Part. I: *Journal of System and Control Engineering*, Vol. 205, No.I4, 1991, pp. 307- 310.

[25] Vaccaro R. J., *Digital control, A state-space approach,* McGraw-Hill, Inc, New York, 1995.

[26] Naumović M. B., PI-like Observer Structures in Digitally Controlled DC Servo Drives: Theory and Experiment, *Electronics*, Vol. 15, No. 1, June 2011, pp. 30-39.

[27] Ogata, K., *Discrete-time Control Systems.* Prentice-Hall, Englewood Cliffs, N.J., 1995.

[28] Naumović M. B., A simple parametric identification procedure based on the final value theorem, *Proceedings of the Institution of Mechanical Engineers,* Part. I: *Journal of System and Control Engineering*, Vol. 226, No. I10, 2012, pp. 1432-1437.

In: Automation Systems of the 21st Century
Editors: D. Arent and M. Freebush
ISBN: 978-1-62948-262-0

Chapter 5

CHALLENGES ON THE WAY TO AUTOMATIC NETWORK INTRUSION DETECTION

***Carlos Catania*[1,*] *and Carlos García Garino*[1,2,†]**

[1]ITIC, Universidad Nacional de Cuyo, Mendoza, Argentina
[2]Facultad de Ingenieria, Universidad Nacional de Cuyo, Mendoza, Argentina

Abstract

The use of Network Intrusion Detection Systems (NIDS) emerges as a tool aimed at helping system administrators to monitor and identify computer attacks. A NIDS monitors network segments and performs analysis at different network protocol layers to identify suspicious activities. In the present chapter we survey the most relevant works in the field of automatic network intrusion detection in the last 15 years. Our goal is to analyze the intrusion detection problem from a wider perspective, which can help to identify the causes behind the lack of acceptance of novel approaches by system administrators as well as provide new lines of work to the intrusion detection research community. In contrast to previous surveys, we consider other issues than just the embedded classification problem. Features such as traffic model acquisition scheme, usage frequency and adaptability are analyzed in two different stages of intrusion

*E-mail address: ccatania@itu.uncu.edu.ar
†E-mail address: cgarcia@itu.uncu.edu.ar

detection. In particular, we considered the initial traffic model generation and the further traffic model adjustment stages. Our research has shown that the level of human interaction is still high during both intrusion detection stages. This situation responds to the fact that some of the assumptions in which these approaches rely on, do not always hold. The availability of network traffic labeled as intrusive or normal, or the presence of attack-free network traffic, are two usual assumptions followed by many alternative techniques. Unfortunately, ensuring such assumptions demands a lot of work from security experts which is precisely what is wanted to avoid.

Keywords: Intrusion Detection, Network Security, Automatic Detection

1. Introduction

Since Internet appearance in the 80s, the number of network computer systems and users have significantly grown. Nowadays, network computer systems have become an important part of our life. It is difficult to imagine our daily activities without accessing at some point to information provided by interconnected computer systems. Internet services such as e-mail, instant messaging and the world wide web (WWW) are not only part of our daily life but also indispensable for normal operation of any business organization. However, the Internet growth along with all its benefits brought with it the rise of new security threats. Vulnerabilities found in Internet protocols such as ARP, TCP, TELNET, SMTP and FTP have caused most of known attacks against network data confidentiality, authenticity and availability. Even though all of these problems are currently fixed, attacks exploiting new vulnerabilities are performed everyday. As announced by the Computer Emergency Response Team/Coordination Center (CERT/CC), the number of incidents reported has evolved from only 6 in 1988 to more a 120.000 in 2003. Whereas vulnerability alerts have raised from 171 submitted in 1995 to more than 7000 in 2007. A few years ago, exploiting vulnerabilities required a deep knowledge on every element of a network computer system. Nowadays, with the appearance of automatic attack tools, exploiting a known vulnerability can be done by anyone with just minor understanding of network operations. In addition, the attacker profile has evolved from the lonely hackers looking for recognition inside their community to large cyber-crime organizations selling their services to unethical corporations, foreign governments or even terrorist organizations.

The process of securing a computer network against these threats has become a difficult task that requires an enormous amount of work. Nowadays, it is very common to have a security engineering staff responsible for securing the complete network. However, everyday new attacks keep exploiting unknown vulnerabilities to gain unauthorized access to the network. Although some of them are just minor adaptations of attacks already known, sometimes it is simply not possible for the security staff to meet the security requirements for defending the network against these new threats.

Clearly, some other security mechanisms are required for improving network security response to present threats. Intrusion Detection emerges as a complementary approach for monitoring and identifying computer attacks. In a recent security recommendation, the National Institute of Standards and Technology (NIST) [88] defines intrusion detection as the process of monitoring the events occurring in a computer system or network and analyzing them for signs of possible incident. At the same time, the list of incidents of interest considered by NIST are violations or imminent threats of violations of computer security policies, acceptable use policies, or standard security practices. }In other words, an incident could be seen as any variation from normal usage according to previously defined security policies. It is worth noting that not only security policy violations are considered as interest incidents but also security policy violation attempts.

A network intrusion detection system (NIDS) is the software tool that automates the network intrusion detection process. From an architectural point of view a NIDS can be analyzed from several angles (i.e. traffic capture process, system location, appropriate measures selection, among others). However, from a more simplified point of view, intrusion detection can be seen just as a classification problem in which a given network traffic event is assigned as *normal* or *intrusive*.

In the past 20 years, several techniques have been proposed to address the embedded classification problem inside NIDS. Perhaps the most successful approach has been the one based on pattern signatures describing known attacks behavior [86]. Under this approach, a malicious event is detected when some monitored event matches against a signature pattern. Despite signature-based NIDS are considered the *de facto* standard, they face the problem of needing a new set of signature patterns each time a new attack emerges. In addition, signatures describing such attacks have to be written by experts, which are not always available. In other words, the signature-based approach has failed in providing the level of automation required by security staff members.

Alternatively, techniques including statistical methods, machine learning and data mining methods have been proposed as a way of dealing with some of the issues regarding signature based-approaches. Such techniques aim at facilitating the work of the network security staff, providing a higher automation in the intrusion detection process along with good detection capabilities. Despite the success in obtaining high accuracy levels, most of these techniques have actually not been deployed in real-life scenarios. This situation suggests that accuracy is not the only goal in the pursuit of automatic intrusion detection.

The present chapter reviews the most relevant network intrusion detection techniques for wired networks, putting special emphasis on the embedded classification problem. However, in opposition to previous surveys on this field, analysis is performed considering not only accuracy results but also other features required for implementing the discussed techniques in real-life scenarios.

The rest of this chapter is organized as follows: Section 2 provides background information about the intrusion detection problem, including attack definitions, a taxonomy and a simplified NIDS architecture. Then, in Section 3, the most relevant approaches applied to intrusion detection are reviewed and compared based on the taxonomy along with common measures related to NIDS. Section 4 remarks the remaining open issues, which aim to explain why all except the signature-based approach are not being deployed on current networks. Finally, concluding remarks are provided in section 5.

2. Background

Before discussing the most relevant approaches to NIDS, we proceed to describe the fundamental elements inside the intrusion detection problem. Intrusion related terminology and definitions are presented in Subsection 2.1. Subsection 2.2 presents the main components of an architecture for a simple intrusion detection system. Finally, a taxonomy for network intrusion detection systems as well as some of the most common measures used for evaluating them are provided in section 2.3.

2.1. Intrusion Definition and Classification

Since the network intrusion detection problem is essentially a classification task where a given network event is identified as *normal* or *intrusion*, it is important to precisely define what it is meant by these words.

A computer *attack* can be defined as the intelligence of evading or evading attempt of computer security policies, acceptable use policies, or standard security practices. In the security research community, the terms *attack* and *intrusion* are often used with the same meaning.

In the past years, there have been several attempts to build taxonomies aimed at classifying attacks. One of the most accepted taxonomies is the proposed by [47], where attacks can be classified into four categories:

- **Probing:** These attacks do not attempt to damage the target system. Instead, the goal is gathering information about the system, probably for being included in further attacks. These attacks includes network traffic sniffing and port/address scans

- **Denial of Service (DoS):** the purpose of these attacks consists of diminishing or totally interrupting the use of a system or a service to their legitimate users.

- **User to Root (U2R):** attacks that aim to gain root access to the system by means of exploiting vulnerabilities in operating systems or software applications. The attacker has a valid account in the system.

- **Remote to Local (R2L):** similar to U2R attacks, but in this case, the attacker does not have a valid account in the system. These attacks are oriented to gain local access from outside.

Since [47] develops its taxonomy considering only those attacks present in DARPA intrusion detection evaluation data sets [62, 63], not all attacks are covered by this taxonomy. In this sense, a broad attack taxonomy is presented in [54], where along with the categories mentioned already above, a new category is added that includes programs that replicate on host machines or propagate through the network. This new category includes programs such as *viruses*, *worms* and *Trojan horses*.

As mentioned by [40], it is difficult to classify attacks considering only one point of view. In that sense, [54] proposes four additional aspects for classifying attacks, which are listed below:

- **The number of connections:** It is possible to classify attacks according to the number of network connections involved in the attack. Typical examples of an attack involving multiple network connections are DoS, probing and worms. On the other hand , in U2R attack such as buffer overflows, just a single or a few connections are involved.

- **Source of the attack:** An attack can be performed from a single location (single source attacks) or from several locations (distributed or coordinated). Large-scale stealthy scans, worm outbreaks and Distributed Denial-of-Services (DDoS) can be mentioned as examples of these distributed attacks type.

- **Environment**: There are several attacks depending on the environment where they occur. For instance, attacks may occur in the network, a host machine or a wireless network among others.

- **Automation level:** Attacks can be categorized according to the level of attack automation. They can be divided as follow:

  - *Automated attacks:* These attacks are based on automatic tools capable of probing, scanning and performing a high variety of sophisticated attacks.
  - *Semi-automated attacks:* a number of scripts for scanning and exploiting system vulnerabilities are deployed and then used by a master to perform different kind of coordinate attacks against the target. In this category we can find zombie networks also called Botnets. According to [56], a **Botnet** is defined as a group of computers infected with malicious programs that operate against owner's intentions and without his knowledge.
  - *Manual attacks:* the whole attack process is performed manually. This typically requires a lot of knowledge and work. These attacks are not very common. However, these are the most difficult to detect.

Many other taxonomies of attacks and vulnerabilities have been published over the years. Moreover, if we consider the complete security field [44] this number significantly increases. Despite this fact, an universally accepted attack taxonomy is still lacking. In the context of this work, we will follow the approach proposed by [47] and [54], since we think both provide a good framework for understanding the nature of computer attacks.

2.2. A Simple NIDS Architecture

Despite the high number of different IDS approaches proposed in the past years, most of these can be described by means of a generalized model. Based on the work of [8], we present a reduced version of an IDS architecture considering only the most relevant modules according to the focus of this work. It is worth noting that despite the model could be a little over simplified, it still provides a good description for understanding the complete intrusion detection process.

A general approach for a NIDS architecture and its operational flow is shown in Figure 1. In a first stage, the traffic data acquisition process is done by a *sensor* deployed in a convenient place in the network, usually the main backbone switch. This *sensor,* continuously monitoring the traffic, gathers all the data available on the wire in raw mode for further processing. Optionally, a *Network Traffic Collector* is available for storing the captured raw traffic.

Then, the *Traffic Features Generator* builds the network traffic features selecting, extracting and rearranging information inside raw traffic packets. In general, a network traffic feature refers to some components of network traffic relevant for incident detection. The set of selected network traffic features composes a traffic record.

Recently generated traffic records are stored in the *Traffic Record Collector.* Then such records are evaluated by the *Incident detector module* using some previously learned *Traffic Model.* When an incident is detected, an *Alert* is raised, and the suspicious traffic record together with information related to the incident are stored for further expert analysis.

According to Figure 1 an architecture for a network intrusion detection system should contain the following modules:

- **Traffic Data Acquisition:** This module is used in the data collection phase. In the case of a Network Intrusion Detection System, the source of the data are raw network frames. However, under some situations, there could be no interest in capturing the complete raw data frames. In those cases, information from upper protocol layers are collected instead (e.g. IP layer).

- **Traffic Features Generator:** Using all the data provided by the sensors in the *Traffic Acquisition Module* not only could be extremely inefficient in terms of both CPU and storage resources but also, in some cases, could lead to an inaccurate detection. It is worth noting that each network traffic feature provide us with a partial point of view of the network where

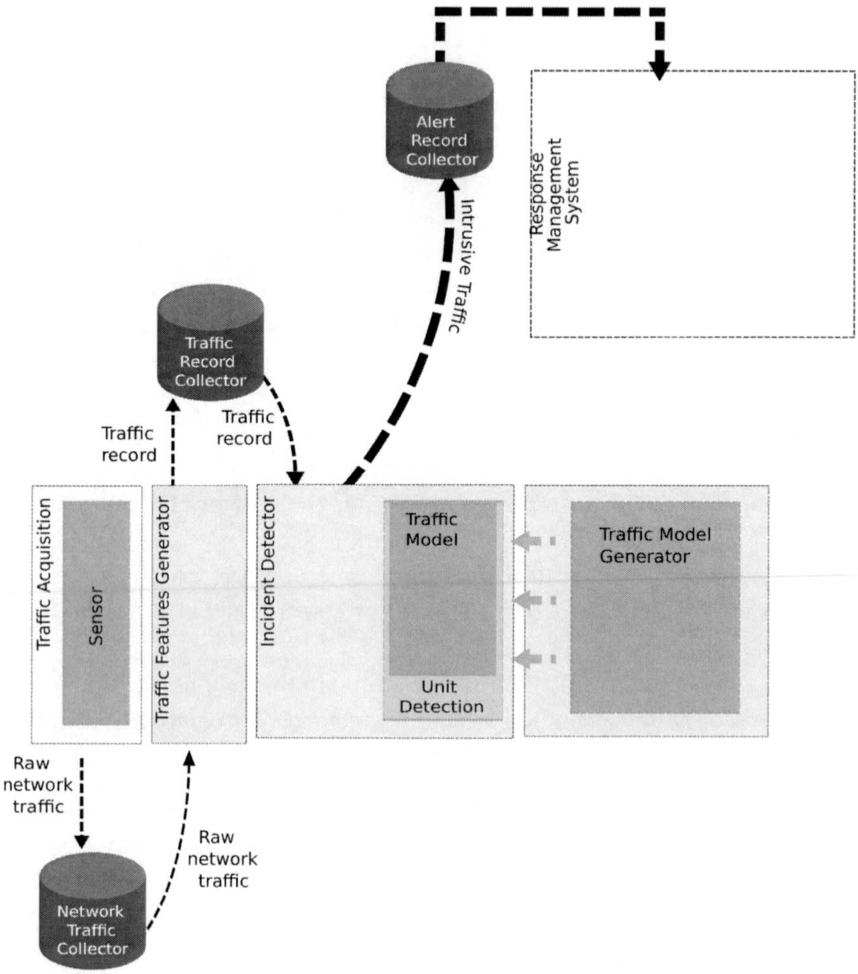

Figure 1. A general NIDS architecture.

not all possible type of attack will be observable. Therefore, it is necessary to formulate explicitly which network traffic features will be in the best interest of the intrusion detection process. The *Traffic Features Generator* Module is responsible for extracting from captured traffic the set of selected traffic features required for an accurate intrusion detection. Network traffic features can be classified in *low-level* features and *high-level* features. A *low-level* feature can be directly extracted from captured traffic (e.g. IP header). Whereas a *high-level* feature consists of traffic information deduced from captured traffic by means of some process. (e.g. average transfer rate). Certainly extracting *low-level* features is less resource-consuming than extracting *high-level* features, an issue that should be considered when the intrusion detection requirements are defined. It is worth noting that each network traffic feature provides us with a partial point of view of the network where not necessarily all possible of types of attacks will be observable. Features can be also classified according to the network traffic source used for generating them. *Packet* features are those directly obtained from network raw packets headers. *Flow* refers to features containing aggregated information related to network connections. Finally *Payload* stands for those features obtained from packet payload.

- **Incident Detector:** This module processes the data generated by the *Traffic Features Generator* module to identify intrusive activities. This is the core component of a NIDS. It is here where the different methodologies for detecting intrusion are implemented. Traditionally, network intrusion detection methodologies have been classified into two broad categories: *misuse detection* and *anomaly detection* [74]. The idea behind misuse detection consists of comparing network traffic against a *Model* describing known incidents. This approach has proved being very effective at detecting known threats but largely ineffective at detecting unknown threats. On the other hand, the anomaly detection method is based on the analysis of the profiles that represent normal traffic behavior. First, anomaly detectors create a baseline profile of the normal "legitimate" traffic activity. Thereafter, any new activity that deviates from the normal is considered an anomaly. This methodology has the major benefit of potentially recognizing unforeseen attacks. However, its major drawback is a potentially high false alarm rate. It is worth noting that not all deviations detected by *anomaly detection* algorithms might be actual attacks, in fact many of these detected anomalies could be new or unusual, but still legitimate, network traffic. [54]. No matter the detection methodology implemented by the *Incident Detector*, once a malicious event has been detected, an alert will be raised and sent to the *Response Management* module.

- **Traffic Model Generator:** Despite of the detection methodology chosen at the *Incident Detector Module*, all of them require some previously acquired reference data (some other authors prefer the term knowledge base instead [54]) to compare against. We can think of this reference data as a model representing traffic behavior, describing normal or well-known event incidents. This reference data is managed inside the *Traffic Model Generator* module. The source of information of the *Traffic Model Generator* could come from human knowledge or from some automatic knowledge acquisition procedures. A deeper discussion about the different approaches implemented in this module is given in Section 3.

- **Response Management:** Once an alert is received, this module has the responsibility to initiate actions in response of a possible intrusion. [8] recognizes two different classes of response behavior: a system is *active* when it automatically takes actions based on information provided by alerts. Some authors often refer to this class of systems as *Intrusion Prevention Systems* (IPS) [88]. On the other hand, in a *passive* systems, alerts are stored for later human operator analysis or sent to a different system for further processing. The fast response of *active* systems has the benefits of potentially stopping possible threats before they become actual intrusions. However, as not all alarms initiated by the *Incident Detector* correspond to actual intrusions, an *active* system may lead to unexpected network behavior. *Passive* system do not perform any changes in network behavior. Their main task consists of summarizing monitored events or providing details on particular events of interest for further human revision. However, if human analysis is not performed regularly, the lack of an immediate response by *passive* systems may result in irreparable damage.

2.3. A Taxonomy for NIDS

In the past years, several authors have proposed a number of intrusion detection system taxonomies. [25] were among the first to propose a taxonomy for intrusion detection. [8] later extends the taxonomy proposed by [25], becoming one of the most cited taxonomies in the field. More recently, [54] have proposed their own taxonomy synthesizing the work of [8] and [25].

Here, we summarize the elements of a taxonomy for Intrusion Detection Systems (IDS) as it was proposed by [54]. Notice that some elements are not included since they are outside the scope of this work and only those elements related to NIDS field are considered. The simplified taxonomy for NIDS considers the following elements:

- **Detection method:** the two broad methods for detection are considered: *anomaly-based* (based on a definition of normal behavior of the target system) and *misuse-based* (matches the input data against a definition of an attack).
- **Model acquisition:** traffic model is generated based on human knowledge or by some automatic generation process.
- **Usage frequency:** detection can be performed in real- time (continuous monitoring) or by batch (periodic analysis).
- **Architecture:** data collection and processing can be done only from a single monitored point of the network (centralized*)* or from multiple points of the network (distributed).

Although the above taxonomy intends to classify NIDS as a whole system, it seems clear that some of the elements show a direct relation to specific modules of the simplified NIDS architecture (section 2.2). For instance, the first two elements of the taxonomy are strongly related to tasks assigned to the *Detection Incident* and the *Response Management* modules, leaving aside the remaining modules. On the other hand, the division between real-time and batch systems encompasses a larger set of modules. Considering a NIDS suitable for performing real-time detection implies adjustments in practically all the modules of the system.

From an architectural point of view, a NIDS can be divided in *centralized* and *distributed.* Logically, these two categories deeply influence the behavior of several modules inside NIDS architecture. For instance, regarding to the *Traffic Acquisition Module,* data collection can be done using only one *sensor* or several *sensors* distributed along the network. In that sense, the *Cisco Netflow* [22] architecture provides a good example of this distributed sensors approach. A graphical representation of this data collection distributed approach is shown in Figure 2

The *Incident Detection* module can also have distributed processing capabilities. A common approach [105] consists of having several *detection units* providing low-level decisions to the *Incident Detector* module. A *detection*

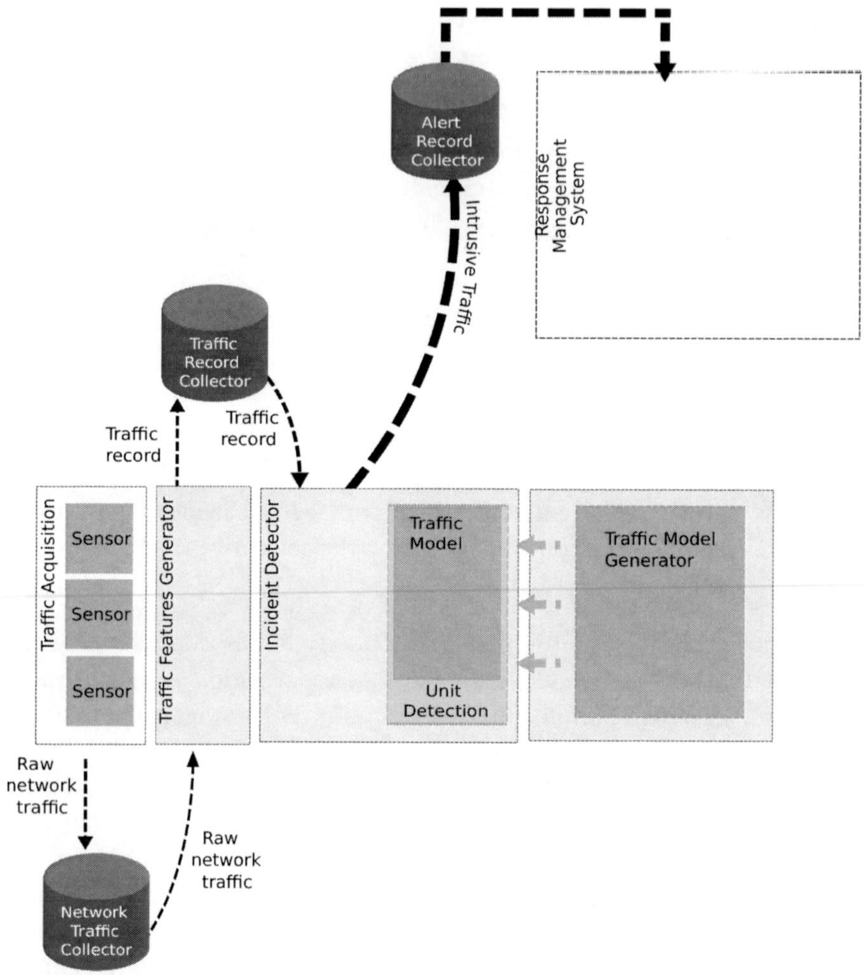

Figure 2. A NIDS architecture with a distributed traffic acquisition module.

unit contain its own (and probably different) *Traffic model*. In a sense, *detection units* can be seen as low-level NIDS by itself. Then, the *Decision Making* module will transform the low-level decisions into a high-level report of confirmed attacks. A graphical representation of the *Incident Detection* distributed approach is shown in Figure 3.

In general, *Incident Detection* distributed approach is focused on using dif-

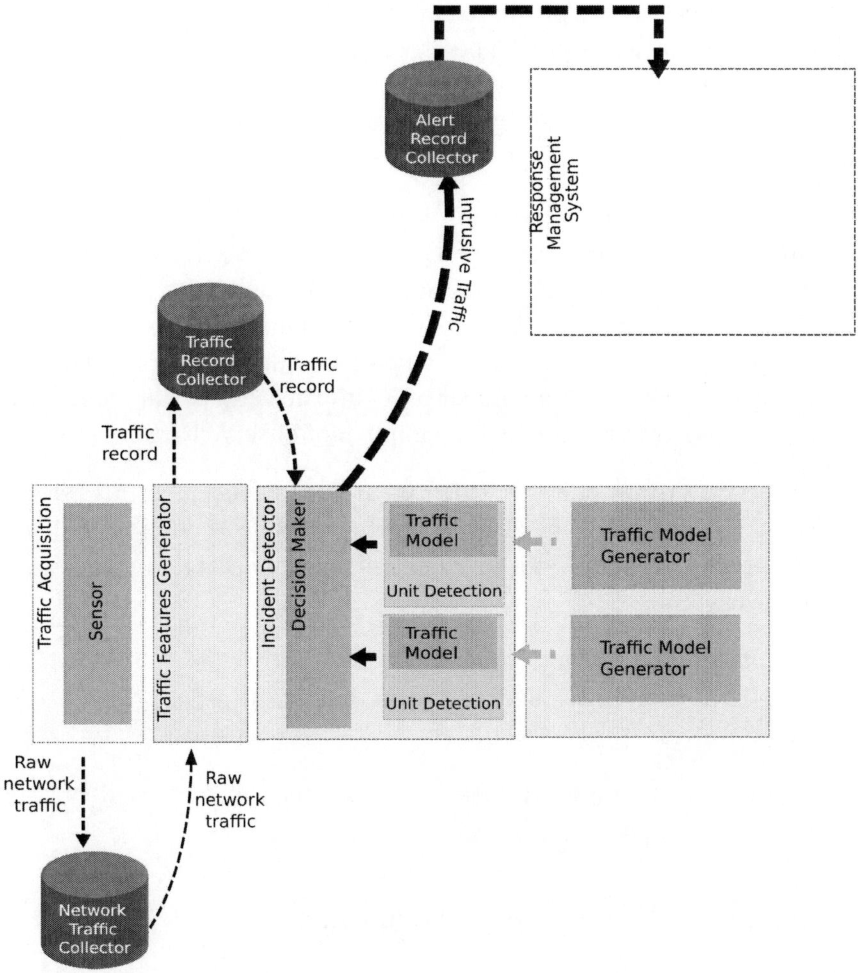

Figure 3. A general NIDS architecture a distributed incident detection module.

ferent classifiers to take a unique decision about the same single network traffic record. Using this distributed approach has the benefits of providing a better representation of the network behavior. However, these approaches bring about some other issues whose solution could require putting additional effort into the already difficult problem of intrusion detection. Such as sensor location, information storage, privacy and information fusion. These and some other distributed approaches are deeply discussed in [105].

Finally, in order to analyze the performance of NIDS, a number of measures should be considered. This issue has been previously considered in [25, 9, 54]. Below, we summarize the four more relevant measures when considering true deployment feasibility.

- **Prediction accuracy:** measures how good is an NIDS in detecting intrusion. Some of the most common performance measures for NIDS evaluation correspond to Attack Detection rate (DR) and False Alarm rate (FA). DR is computed as the ratio between the number of correctly detected attacks and the total number of attacks. While the FA rate is computed as the ratio between the number of legitimate actions that are incorrectly classified as attacks and the total number of observed actions.

- **Processing time**: considers the rate at which events are processed. An NIDS should be able to perform detection as soon as possible. Otherwise, the security staff could not have enough time to react before much damage has been done.

- **Adaptability:** attacks techniques are constantly changing over the time. An NIDS should be able to re-adapt itself fast in the presence of new attacks.

- **Resource consumption:** measures how much memory and storage resources are required by the system.

3. Intrusion Detection Approaches

This Section reviews the more relevant works in the area of NIDS. More than 100 research works have been surveyed for writing this chapter. However, since discussing them all results in an infeasible task, we have focus the analysis on those works with the highest impact in terms of number of citations. Nevertheless, the remaining ones (i.e. those works with lower impact) have been included in the tables shown at the end of this Section.

Following the simple NIDS architecture presented in section 1 in the previous Section, we observe that most of the work done by the intrusion detection research community has been focused on the *Traffic Features Generator* and *Incident Detection* modules. Therefore, the focus of the analysis is carried out on

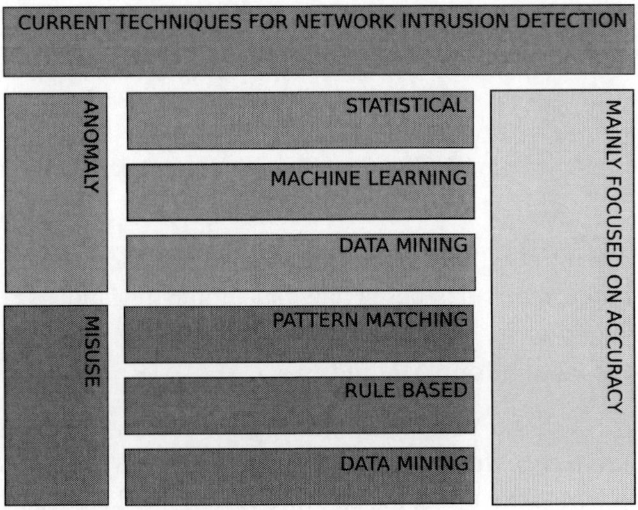

Figure 4. Intrusion Detection methods categorized by technique.

the *Incident Detection* module of an NIDS, including details about the *Traffic Model Generator* along with some brief information about the features used by each approach.

Because of the large number of works presented during the past years for both misuse and anomaly detection, it seems convenient to group them according to the techniques used by each one of them. In this sense, we rely on the categorization proposed by [76] and [54] and we cluster each approach as shown in Figure 4.

The next section reviews the most significant research works based on the misuse detection approach. Section 3.2 provide detailed analysis about most relevant approaches based on anomaly detection.. Then, in Section 3.3 a deep analysis about the benefits and flaws regarding each approach is presented.

3.1. Misuse-Based Intrusion Detection

As mentioned in section 2.2, there are two methodologies for intrusion detection: *misuse detection* and *anomaly detection* [74]. The idea behind misuse detection consists of comparing network traffic against a *Model* describing known intrusion events. This approach has proved to be very effective at detecting known threats but largely ineffective at detecting unknown threats. Among

```
alert tcp any any -> 10.1.1.0/24 80
(content: "/cgi-bin/phf"; msg: "PHF
probe!";)
```

Figure 5. A simple SNORT pattern signature. Adapted from [86].

the most commonly used techniques for misuse detection we can find pattern recognition, implication rules-based and data mining techniques.

3.1.1. Pattern Recognition Methods

Nowadays, the most successful strategy for intrusion detection is the so called *signature-based.* In this approach each malicious event is described by means of a pattern signature using its own descriptive language. A malicious event is detected when some monitored event matches against this signature pattern. SNORT is a lightweight NIDS originally developed by [86] in 1999 that has become the facto standard for intrusion detection. It is a complete and fully functional IDS that follows a signature-based approach.

The SNORT architecture, described in the work of [86], is focused on real-time performance, simplicity and flexibility. It is worth noting that SNORT also follows the general architecture presented in Section 1. In that sense, three modules can be identified inside SNORT: a *packet decoder*, a *detection engine* and a *log/alert subsystem*. The *detection engine,* which resembles the *Incident Detector* module in the general NIDS architecture, contains the pattern signatures for security incidents. These signatures are then searched for each packet. The first signature that matches a decoded packet in the detection engine triggers a previously specified action and returns. Clearly in the case of SNORT, the source of the *Traffic Model Generator* comes from the human knowledge described by the thousands of pattern signatures included into SNORT database. These signatures are able to perform content pattern matching and detect a variety of attacks and probes, such as buffer overflows, stealth port scans, CGI attacks, and more. An example of SNORT pattern signature is shown in Figure 5. In this case, the rule would detect attempts to access the PHF CGI script on any of the local network's web servers.

SNORT requires that the signature database be regularly updated by the security staff in order to recognize new security threats. Alternatively, SNORT provides a way to adapt itself to new intrusion events automatically updating its signature database from a central repository.

The signature approach has been included into other NIDS. Bro [78] is a well-known NIDS focused on real-time operation that follows a signature-based approach. SURICATA [95] is another signature-based NIDS recently launched by the Open Information Security Foundation (OISF). SURICATA claims to be a replacement to the 12-year old SNORT, providing improved multi-thread support for multi-core architectures. Despite the differences related to performance and efficiency, all of these NIDS have support for including in some way the SNORT signature database.

3.1.2. Implication Rules-Based Methods

Close to the pattern matching approach discussed in the previous section, is the idea of using a set of implication rules to describe an intrusion event. Basically, in this approach the *Incident Detector* module of the NIDS is implemented by means of a rule-based expert system. [61] discuses P-BEST, a rule-based forward chaining expert system as a part of the EMERALD NIDS [81]. The idea behind this approach consists of giving a set of rules describing network events, a series of these asserted network events may lead the system to infer that a previously defined intrusion event has occurred. In their work, [61] provide an usage example of the P-BEST inference engine for detecting DOS attacks. Authors also provide performance measurements indicating that P-BEST-based expert systems are potentially well suited for real-time misuse detection.

3.1.3. Data Mining Techniques

In their much cited work of 1998, [55] proposed a framework for applying data mining techniques to the NIDS field. Their research aimed to eliminate the need of manually created *Traffic Models* by automatically building them from some reference data.

Data mining algorithms for misuse detection require traffic records labeled as *normal* or *intrusive* in order to build a *Traffic model.* Major benefits provided by these algorithms consist of being able to detect not only known attacks but also their variations

One of the first data mining approaches was proposed by [18]. On his work, he proposes a misuse approach different from the rule-based approaches proposed so far. The work is focused on a description of an NIDS based on Artificial Neural Networks (ANNs).

The configuration proposed by [18] follows the general architecture shown in Section 2.2. However, only the *Incident Detector* and *Traffic Features Generator* modules are considered. The ANNs working as the *Incident Detector Module* would receive data from the *Traffic Features Generator* module and analyze the information for instances of misuse. Any instances that are identified as indicative of attack would be forwarded to the security staff or used by an automated intrusion response system.

The *Traffic Features Generator* module receives data from the network and extract only nine low-level features from each packet, as can be seen in Table 1. According to [18], these nine features provide a complete description of the information transmitted by each packet.

Table 1. Features provided by the *Traffic Features Generator* Module of the NIDS approach proposed by [18]

Protocol ID	The protocol associated with the event
Source Port	The port number of the source.
Destination Port	The port number of the destination
Source Address	The IP address of the source
Destination Address	The IP address of the destination.
ICMP Type	The type of the ICMP packet
ICMP Code	The code field from the ICMP packet
Raw Data Length	The length of the data in the packet.
Raw Data	The data portion of the packet.

In a second part of the feature generation phase, three of the nine data elements (ICMP Type, ICMP Code and Raw Data) are converted into a standardized numeric representation for the further processing by the *Incident Detection*

Module. Following a common strategy in misuse-based detection approaches, a tenth element is assigned to each record whether the event represented is part of an attack on the network (a process usually referred as *labeling*).

The ANNs inside the *Incident Detection* Module utilizes a feed-forward Multi Layer Perceptron (MLP) architecture that consists of four fully connected layers with nine input nodes and two output nodes.

As mentioned by [18], despite not being a completely functional NIDS, the two modules implemented are enough for showing the viability of the whole ANN approach.

Regarding the ANNs approaches, a functional misuse NIDS based on ANNs is proposed by [14], in their work they present a complete architecture for NIDS where ANNs are representing the suspicious state of a particular connection based on previous well known intrusion patterns. It should be noted that both ANNs approaches [18, 14] require model adjustments for reflecting new intruders behavior patterns. In this case, new intrusion patterns need to be added to the network that must be retrained in order to learn to recognize the new patterns appropriately.

The use of ANNs for misuse detection has been one of the most analyzed approaches for IDS. Since the work of [18], many other authors have followed this line of research. Among the more relevant, we can cite the work of [62]. In their work, the authors analyze the performance of neural networks on *U2R* intrusion events by looking for pre-defined keywords over user sessions. The work done by [75] is another important contribution to the field. In their work of 2002, the authors focused on discovering a set of relevant features that describe user behavior on a system, and then used the set of relevant features to build ANNs classifiers. We can also cite the work of Liu et al. [64], in which they applied Principal Component Analysis (PCA) combined with ANN for automatic feature extraction. More recently, Kumar et al. [49] have applied ANN using the resilient back propagation algorithm for detecting distributed DoS attacks. A summary of most of the research done on ANN applied to the NIDS field in the last 20 years can be found in Ahmad et al. [6].

Evolutionary computation (EC) is another data mining approach that has played an important role in the IDS research field due to its adaptive characteristics. In particular, the use of genetic algorithms (GAs) [42] has raised a considerable interest by the intrusion detection research community.

GAs have been applied in several tasks regarding NIDS, such as automatic model structure design [100, 58] and the generation of function transformation[21, 5]. However, GAs have been mostly used for building rule-based models for classification [59, 36, 12, 96]. Instead of manually writing the rule set, as needed by traditional rule-based system, the idea of using GAs for classification is to automatically obtain such classification rules. The main advantage of rules is that they are simple and easy to understand.

The works of [59] and [36] give the implementation details of a simple NIDS using GAs for evolving classification rules. We proceed to discuss their work in terms of the general architecture from Section 1.

In their work both authors describe the elements behind the *Traffic Features Generator* and the *Incident Detection* modules. The *Feature Generation* module implemented by [36] considers the connection time (a high-level feature) and five low-level features from the network packets and TCP connections: protocol type, source and destination port, source and destination IP address and the label of the traffic record. [59] extends the number of features to ten, adding information about the connection state and other high-level features such as the number of bytes transferred from source and destination addresses, respectively. These features are then used as the condition part of the evolved rules. An example is shown in Figure 6. Evolved rules could then be used as part of traditional rule-based systems such as the P-BEST [61] system mentioned previously.

```
if {the connection has following information:
source IP address 10.0.1.1; destination IP
address:  192.168.176.2; source port number:
42335; destination port number:  80; connection
time:  482 seconds; the connection is stopped
by the originator; the protocol used is TCP;
the originator sent 7320 bytes of data; and the
responder sent 38891 bytes of data } then {possible
intrusion}
```

Figure 6. Evolved rules describing misuse behavior. Adapted from [59].

Although both authors implement a simple GA inside the *Incident Detection* module, their algorithms differentiate in their fitness functions. [59] calculates the overall outcome based on whether a field of the connection matches the pre-classified data, and then multiply the weight of that field. The matched value is set to either 1 or 0. On the other hand, [36] uses the support-confidence framework [98]. Both approaches were tested on the DARPA intrusion detection evaluation data sets [62, 63]. Experimental results showed that the proposed method works effectively for the selected data sets.

As mentioned in the work of [35], the normal and the abnormal behaviors in networked computers are hard to predict, as the boundaries cannot be well defined. This has lead some authors [35, 16, 34, 3, 20, 4] to suggest the quality of the rules obtained by simple GAs can be improved by means of combining GAs with fuzzy logic. In this sense, [34] used an adaptive-parameter GA with special operators such as gene addition and gene deletion for evolving fuzzy rules describing normal and intrusion behavior. Clearly, the model obtained using evolved fuzzy rules gives a better description of the network behavior. However, it is worth noting that the complexity added into the *Traffic Model Generator* can translate into larger building time requirements.

Beyond the use of ANNs, GAs and fuzzy logic, many other data mining approaches have been applied to the intrusion detection field. Some authors [66, 29] preferred using fuzzy association rules, avoiding the large building time requirements of the fuzzy GAs approaches. [24, 65, 5] applied Genetic Programming (GP) algorithms, another EC approach similar to GAs, but with different implementation details. [101, 91] proposed an approach for signature recognition based on decision trees. [75] analyzed the behavior of Support Vector Machines (SVM) for recognizing intrusion patterns [75, 94]. In a similar fashion, [102] proposes an approach to improve incremental capabilities of SVM that are necessary for performing real time detection. Swarm intelligence [13] is another technique that has been deeply analyzed regarding the network intrusion detection problem. In this sense, Ant Colony Optimization (ACO) has been analyzed in the works of [84, 11, 41, 1]. On the other hand, Particle Swarm Optimization (PSO) techniques have been applied in the work of [39].

To conclude, special attention should be given to multiple classifiers approaches [48]. From a architectonic point of view, these approaches have an *Incident Detection* module with distributed classification capabilities as shown in Section 2.3. The claimed benefits of these multiple classifiers approaches are that by combining the opinions of an ensemble of classifiers, the new opinion will be better than the individual ones. Recently, the intrusion detection community has started to consider the viability of this compound approach. In this sense, perhaps one of the most relevant study on this topic is the work of [33]. Here the authors analyze the behavior of a multiple classifier approach using multiple ANNs *Traffic Models* generated from a different number of *Traffic Features*. In addition, five fusion strategies are tested for combining the output of the different models.

3.2. Anomaly-Based Intrusion Detection

The anomaly detection method is based on the analysis of the profiles that represent normal traffic behavior. First, anomaly detector creates a baseline profile of the normal "legitimate" traffic activity. Thereafter, any new activity that deviates from the normal model is considered an anomaly. Several techniques have been used for building the baseline profile of legitimate traffic, and many others techniques for recognizing deviations from defined normal behavior as a potential anomaly. This methodology has the major benefit of potentially recognizing unforeseen attacks. However, its major drawback is a potentially high false alarm rate. It is worth noting that, as suggested by [54], not all deviations detected by anomaly detection algorithms might be actual attacks, in fact many of these detected anomalies could be new or unusual, but still legitimate, network traffic. Among the most commonly used techniques for anomaly detection we can find statistical methods, machine learning and data mining techniques.

3.2.1. Statistical Methods

Some of the earliest approaches for anomalies detection were based on statistical methods. The idea behind these methods consists of monitoring the network traffic activities and building a profile to represent their behavior. The profile is based on measures of certain variables over the time. Such measures include the traffic rate, distribution of packets over protocols, the rate of connections, the number of different IP addresses, etc. Usually, these methods require maintaining two profiles during the anomaly detection process: the currently observed profile and the previously stored statistical profile. As a new network event is observed the current profile is updated and compared against the stored profile. An anomaly score is then obtained indicating the degree of irregularity for an specific event. If the score surpasses certain threshold, then this event will be considered anomalous by the system.

The Network Intrusion Detection Expert System (NIDES) [7] developed at the Stanford Research Institute (SRI) is one of earliest IDS. NIDES contains an *Incident Detector* Module based on statistical anomaly detection. NIDES originally efforts were oriented toward the surveillance of user-session and host-layer activity. However, the incident analysis methodology was extended to support network traffic anomaly detection [80] and later included into the EMERALD

Table 2. Basic features provided by the *Traffic Features Generator* Module of the NIDS approach proposed by [80]

Categorical	**Source address**	The IP address of the source
	Destination address	The IP address of the destination
	Protocol	The protocol associated with the event
	Malformed Service	Track the occurrence of malformed request
	Malformed Packet	Track the occurrence of malformed packet
	File handle	Track the access to unauthorized files
	Command	The commands issued (in telnet, ftp sessions)
Continuous	**Number of Bytes**	The number bytes transferred
	Numbers of Packets	The number of packet transferred

NIDS [81]. The statistical module of EMERALD is focused on providing real-time surveillance of TCP/IP-based networks for malicious or exceptional network traffic.

The *Feature Generation* Module inside EMERALD tracks network traffic via four classes of features: categorical, continuous, intensity, and event distribution. Categorical measures are basically low-level features that assume values from a categorical set, such as originating host identity, destination host, and port number. Continuous measures are high-level features for which observed values are numeric or ordinal, such as number of bytes transferred. Table 2 shows the categorical and continuous features used inside EMERALD statistical *Incident Detection* Module.

In addition to basic features, the *Traffic Features Generator* module also includes a set of derived features for tracking the intensity of activity (that is, the rate of events per unit time) and the "meta-distribution" of the measures affected by recent events.

These measures are then used by the *Traffic Model Generator* Module for building the short-term and long- term behavior profiles based on the short-term and long-term probability distributions for each measure. Then, *the Incident Detector* Module compares the short-term profile against the long-term profile, and the difference between the two is obtained. The degree of difference between a long-term profile and the short-term profile for each measure is quantified using a chi-square-like statistic, with the long-term profile playing the role of the actual probability distribution and the short-term profile playing the role of the observations. Later, this difference is evaluated against a historically adaptive distribution of deviations. The empirical distribution of this deviation is transformed to obtain a score for the event. Finally, those events whose scores exceed a historically adaptive score threshold (also based on an empirical score distribution) are considered anomalous. A detailed description of the algorithm can be found in [45].

SPICE (Stealthy Port Scan and Intrusion Correlation Engine) [92] is another statistical based anomaly detection approach focused on detecting slow randomized scans that are buried in days, weeks or even months of normal traffic. The approach followed by SPICE is different to previous approaches that use time traffic windows (looking for at X attempts over Y period of time). SPICE is provided as a SNORT [86] plug-in and is capable of detecting stealthy scans in real-time.

SPICE consists of an anomaly sensor and a correlator. The sensor monitors the network and assigns an anomaly score to each event. The anomaly score for a packet is based on a frequency-based mechanism. The fewer times an packet is observed the higher its anomaly score will be.

Only 4 low-level features are considered by the SPICE *Traffic Features Generator* module: source and destination IP addresses and source and destination ports. SPICE contains a probabilities table that maintains the occurrences of different events along history. The probabilities are calculated using the joint occurrence of the different values of the 4 features. An estimate of the joint probabilities based on Bayes networks is used as a way to avoid to calculate all the different combinations of values while maintaining computing time low. In addition, the network history is weighted attempting to of provide more importance to more recent packets.

Finally, those events above a previously defined threshold, are considered anomalous and passed, together with their anomaly scores, to the SPICE correlator engine. This correlator groups events together and reports scans.

Despite SPICE has been deeply used due to its easy interaction with SNORT [86] NIDS, its very high false alarm rate is well known. This behavior can be explained since SPICE gives high scores to any unseen event despite it is a real intrusion or not.

Some other interesting statistical approaches have been proposed as a way to deal with the non-stationary propriety of network traffic. In this sense, it is important to mention the work of [52]. In the their work of 2004, an exploratory Principal Component Analysis (PCA) is conducted on flow aggregation, namely on origin-destination flows between Points of Presence (PoP) of two European networks. The authors found that backbone traffic can be decomposed in three types of *flows*: *deterministic flows* that observed a periodical tendency (day-night pattern), *spike flows* that show isolated values clearly deviating from the average and *noise flows* that according to the authors appear to be roughly Gaussian. In further works [50, 51], several statistical anomalies detection methods have been applied over the spike components. The proposed methods seem to be appropriate for detecting DoS, scans and worms attacks. On the other hand, [57] propose the use of some information-theoretic measures, such as *entropy* and *information gain*, as a way to analyze the behavior of different anomaly detection methods. These metrics can provide valuable information of audit data and help in parameter setting and model building. Recently, these metrics have been used by [38] as a way to distinguish anomalies that change the traffic either abruptly or slowly. Alternatively, [73] have started to analyze the consequences of applying maximum entropy detectors under different network sizes and network authorities.

3.2.2. Machine Learning Techniques

The use of Machine Learning (ML) techniques for anomaly detection has been also considered by the intrusion detection research community. The applicability of ML techniques shares many principles with the statistical methods. However, they have some differences. Statistical methods put their focus on understanding the process that generates the data. On the other hand, ML techniques are focused on building a model that improves and adapts its performance based on previous results.

One of the most remarkable efforts in the study of anomaly detection is the work of [72]. They have proposed several anomaly detection models based on ML techniques [71, 70, 72]. One of the first approaches proposed is the PHAD system (Packet Header Anomaly Detector) [71]. PHAD performs anomaly detection using previous information from packet headers. We proceed to describe

PHAD according to the general architecture discussed in Section 2.2, PHAD *Traffic Features Generator* module considers 33 low-level features based on fields from the Ethernet, IP and transport layers (TCP, UDP and ICMP). [71] propose a method for dealing with the non-stationary property of network traffic. For *Traffic Model Generation,* PHAD is first trained with attack-free network traffic data. During this stage each feature can have many possible outcomes. For each feature x a probability $1/P(x) = n/r$ is computed. Where n is the times a packet field has been observed and r the number of distinct values. Unlike SPICE and EMERALD whose models adapt to changes by exponentially decaying the training count to favor recent events, PHAD only considers the more recent novel events and discards the rest. Thus, for each feature with a value never observed during training, the *Incident Detector* Module calculates anomaly score for the feature using $t/P(x) = tn/r$. Where t is the time since the last observed anomalous event. Finally, the sum of the scores from all the features is then assigned as the final anomaly score of the event.

ALAD (Application Layer Anomaly Detection) is another approach proposed by [72] that follows the same strategy for dealing with non-stationary process learning. But instead of analyzing single packets, it considers incoming server TCP connections. The *Features Generation* module considers low-level features such as sources, destination addresses and ports of the TCP connections, as well as TCP flags and information from the first word of the TCP payload. Different combinations of these features are considered for calculating the joint and conditional probability occurrence of values within a packet header. LERAD (LEarning Rules for Anomaly Detection)[68] is similar to the ALAD approach, but instead of using a fixed set of probabilistic rules, LERAD replaces them with an algorithm for selecting them based on the training data.

3.2.3. Data Mining Techniques

Data mining techniques have also been used for anomaly detection. [55] propose the use of inductive rule generation algorithms. These algorithms combine the application of association rules with frequent episode patterns to classify network traffic. The *Feature Generation* Module described in [55] has the following low-level features based on packet information: start time, participating hosts, ports, flag protocol. High-level features are considered as well, including the statistics of the connection (e.g., bytes sent in each direction, resent rate, duration, among others). The *Feature Generation* Module also considers some continuous and intensity features such as the number of connections establishment errors or the per-connection average duration of all connection in the past n seconds.

Then, using RIPPER [23], a well-known rules induction algorithm, the model is generated from attack-free network traffic data and using its destination port as class label. Finally, during the *Incident Detection* stage, any new event not matching against any of the learned rules is considered an anomaly.

Alternatively, to avoid the need of attack-free data, some authors have proposed the use of unsupervised learning techniques. These techniques are able to generate a model using unlabeled data as input. Unsupervised anomaly detection algorithms make two assumptions about the data. First, the number of normal instances vastly outnumbers the number of intrusions and second, the intrusions themselves are significantly different from the normal instances. The basic idea is that since the intrusions are both different from normal and rare, they will appear as outliers in the data.

[82] propose an unsupervised approach that uses clustering techniques applied to the intrusion detection problem. Clustering techniques consist of grouping data into clusters according to some distance or similarity measure. In their work, the authors discuss in detail the *Features Extraction* and *Incident Detection* modules.

The features generated from network traffic are similar to the ones used by many of the previously discussed works. They include the basic low-level features of an individual connection as well as other high-level features obtained using some domain knowledge, such as the number of file creation operations, number of failed login attempts, and others. In addition, a number of high-level features computed using a two-second time window are considered. These time-based features provide context information about a connection such as the number of connections to the same service as the current connection or the number of connections to the same host as the current connection, just to mention a few of them. Finally, since every feature is on different scale, a normalization process is carried out. The *Traffic Model* generation follows a single-linkage

[19] clustering approach. The algorithm starts with an empty set of clusters, and generates the clusters with a single pass-through the data set. For each new network event retrieved from the training set, it computes the distance between it and each of the centroids of the clusters in the cluster set so far. The cluster with the shortest distance is selected, and if that distance is less than some constant W (cluster width) then the instance is assigned to that cluster. Otherwise, a new cluster is created with the network traffic event as its center. Then, traffic events are labeled following the assumption that the number of normal events exceeds the number of intrusions, a percentage N of the largest clusters are assumed to contain only normal traffic events while the remaining clusters only contain intrusion events. Finally, during the *Incident Detection* stage, any new traffic event is classified according to the label of its closest cluster.

The distance measure between traffic events plays an important role in clustering algorithms. In their work, [82] propose the use of the standard *Euclidean metric* distance with equally weighted features that is certainly the most common approach. However, some others authors [53] have proposed the use of the *Mahalanobis* distance [67], in which the contribution of each feature is weighted according to the variability of each variable. It is worth noting that distance is not the only measure for computing cluster membership. Some other techniques compute the relative density of each event to compute its membership. In this sense, the Minnesota Intrusion Detection System (MINDS) [27] has an anomaly-based *Incident Detection* module that assigns an anomaly score to a given traffic event based on a density measure known as Local Outlier Factor (LOF). The LOF score is calculated as the ratio of average local density of the k nearest events of a given traffic event and the local density of the event itself.

The use of Support Vector Machines (SVM) for anomaly detection is another interesting technique applied to unsupervised anomaly detection [97, 28]. The SVM variant for anomaly detection consists of finding an hypersphere with the minimal radius R and center c that contains most of the traffic data mapped into a feature space by some kernel function. One of the major advantages of using SVM techniques for anomaly detection is their capability handling data with not only normal traffic but also anomalies. However, SVM techniques relies on the same two assumption as clustering: intrusions are both different from normal and infrequent.

Several more anomaly detection based on data mining techniques have been applied to intrusion detection. Among others, we can mention the use of fuzzy logic [26], parallel genetic algorithms [2] or Artificial Neural networks [83].

Finally, it is worth mentioning that multiple classifiers approaches have also been applied for anomaly detection [15, 90, 85, 79, 32]. In [32], the authors propose an unsupervised Multiple Classifier System (MCS). Each unsupervised classifier is used for modeling a particular group of similar protocols or network services. The use of a modular MCS allows the security staff to choose a different traffic model and decision threshold for different groups of network services. This also allows the security staff to tune the false alarm rate and detection rate produced by each module to optimize the overall performance of the ensemble. The work of [85] is another relevant approach that applies multiple classifiers. In this case five well-known anomaly detection algorithms are combined by means of a *trust modeling framework [87]* used to assign proper trustworthiness and reputation to each traffic model.

3.3. Discussion

Tables 3 and 4 summarize the main characteristics of the intrusion detection approaches described in the previous sections.

Proposed intrusion detection approaches are arranged by row. Each row describes the most representative approach of its kind. The first three columns refer to some of the NIDS taxonomy categories along with the measures discussed in section 2.3. The first column symbolizes the techniques used for generating the model in each work. The second column denotes whether the proposed approach is capable of dealing with traffic detection in real-time. The third column refers to the potential adaptability to new intrusions observed by each approach. Notice that adaptability is measured considering the less possible human interaction into the model acquisition processes as well as the periodic model adjustment caused by the occurrence of new threats. The fourth column alludes to the kind of features used by each approach. Features are cataloged according the classification previously mentioned in the *Traffic Features Generator* module discussed in Section 2.2. The fifth column refers to the kind of attacks potentially detected by each approach. The recognized attacks are mentioned according to the attack taxonomy presented in Section 2.1. Finally, in the last column, other works following the same approach are referenced.

3.3.1. Misuse-Based Approaches Analysis

Table 3 describes approaches following the misuse detection method. Misuse has been the most successful approach for intrusion detection so far. There have been some fully implemented NIDS available in the market for more than 10 years and their benefits have been proved beyond doubt.

At first (probably owing to the tendency of the 90s), most of misuse-based NIDS were basically expert-systems like P-BEST[61]. Despite their initial success, as the number of different attacks grows, it quickly became clear that these approaches were not suitable to perform real-time detection. Since the reasoning algorithms inside expert-systems are extremely computer-intensive, as the rule set gets bigger the system becomes slower and eventually turns unusable.

Alternatively, SNORT [86] proposes the use of signatures for *misuse* detection. SNORT performs packet inspection using fast pattern-matching algorithms. In some way, this signature-based approach emerges to solve the lack of detection speed for processing network events that rules-based system had shown. SNORT is capable of full real time detection when deployed in network up to 1Gbps. However, it is worth noting that current network throughput has overcome initial signature-based capabilities. SNORT and even Bro [78], have exhibited high resource consumption when confronted with the overwhelming amount of data found in today's high-speed networks.

Both SNORT and P-BEST deeply rely on human activity for their traffic model acquisition process. Every time new attacks emerge, a set of rules (or signature patterns) describing such attacks have to be written by experts. This need of human experts results in poor adaptability to deal with new threats. First, when a new kind of attack emerges, the answer time of human experts could be not fast enough to adapt the system to this new threat. Second, it is practically impossible for human experts to be aware of all possible emerging attacks. This could lead to situations where those systems will simply be unable to detect new intrusions.

Misuse data mining techniques [55, 18, 59, 34] arise with the aim of providing a way to reduce the need of human experts for building and adjusting the model. These systems are able to adapt their models to new security threats, if they have access to the network traffic containing such event. Using previously seen network traffic helps to build more accurate models with potentially more capacity for adapting themselves to new kinds of attacks. However, it should be noted that all misuse data mining approaches are unable to detect any attack type not previously seen.

There are also some other major issues to be considered when using misuse data mining techniques. First, data used for building traffic model ought to be large enough for being representative of network traffic.

Table 3. Comparative analysis of Misuse-based Intrusion detection approaches.

Authors	Model Generation Technique	Usage frequency	Model Acquisition/Adaptability	Features	Attack Type	Related Works
Lindqvist and Porras (1999)	implication rules	real*	**human:** rules written by experts	**low and high-level:** packet, payload, flow	DoS, R2L	[81]
Roesch (1999)	pattern signature	real**	**human**: signature patterns written by experts	**low-level:** packet, payload	R2L, U2R	[95, 78]
Cannady (1998)	**Data mining:** ANNs	batch	**automatic**: retraining with new attack patterns	**low-level**: packet	DoS	[46, 104, 49, 64, 75]
Li (2004)	**Data mining:** GA	batch	**automatic**: retraining with new attack patterns	**low and high-level:** packet, flow	DoS, Probe	[96, 12, 36]
Gomez et al. (2002)	**Data mining:** fuzzy/GA	batch	**automatic**: retraining with new attack patterns	**low and high-level:** packet, flow,payload	DoS, Probe, U2R, R2L,	[4, 20, 5, 16]

0.5 (*) Real time performance only with a reduced set of rules

(**) Real time performance under some bandwidth scenarios

Certainly, since network traffic follows a non-stationary behavior, obtaining such representative sample cannot always be guaranteed. Second, for meeting building model requirements, network traffic data have to be previously labeled as *normal* or *intrusive*. This labeling process of thousands of traffic records is a very time consuming task that demands a lot of expert knowledge. Finally, since the building model processes of data mining algorithms are highly computer intensive, most of them are certainly not capable of performing real-time detection.

3.3.2. Anomaly-Based Approaches Analysis

Table 4 describes those approaches based on anomaly detection. The interest of NIDS relying on anomaly-based *Incident Detection* modules has increased considerably in the recent years, mostly because their ability to detect forms of intrusions never seen before.

Statistical-based approaches [92, 50, 80] have the benefits of not requiring prior knowledge of attacks for their model generation process. Most of the statistical-based approaches have proved to be suitable for real-time detection. In this regard, approaches proposed by [80] and [92] have been tested and have shown an acceptable performance in real traffic situations. The work of [52] has not been fully implemented. However, in their work, the authors provides experiment details that seem to prove the validity of the approach for real-time detection.

Like in all the anomaly detection approaches, in the case of statistical-based, adaptability does not mean to adjust the model for detecting new kind of intrusions. Instead adaptability refers to the required adjustments in the *normal* traffic model each time the network traffic behavior changes. Such adjustments are done using attack-free network traffic data.

Statistical-based approaches assume that network traffic responds to a quasi-stationary process. A situation that is not always realistic [30]. This incorrect assumption is perhaps the major drawback regarding the adaptability of statistical-based approaches and one of the causes behind the high false alarm rate reached by these methods.

On the other hand, the use of ML and data mining techniques in the intrusion detection field has been beneficial due to their potential adaptability to changes as new information is acquired. For instance, the approach proposed by [69] has the major advantage that no distribution assumption is made during the traffic model generation, a situation that facilitates the automatic adaptation process.

Despite their potential improvements, data mining (and also ML) approaches for anomaly detection share some of the issues of misuse-based data mining techniques. First, since they have a computational intensive model generation process, these techniques have been considered mostly for batch detection. Second, they need a large amount of network traffic data labeled as normal for the model generation process. However, there is a subtle difference regarding labeled traffic requirements. Misuse-based data mining techniques assume the availability of fully labeled data, whereas anomaly-based just require attack-free data. This last approach is usually referred as *supervised anomaly detection* [19]. At first glance this could look as a improvement over misuse approaches. In practice, however, it is difficult to obtain attack-free data to implement these approaches. Verifying that no attacks are present in the training data can be an extremely demanding task, and for large samples this is simply infeasible. On the other hand, if the data containing attacks is treated as clean, intrusions similar to the ones present in the training data will be accepted as normal patterns, resulting in a increment in the number of misdetections.

Unsupervised anomaly detection approaches [82, 27, 53, 28] arise as a way to deal with the *supervised anomaly detection* limitations regarding labeled traffic requirements. [82] propose the use of unsupervised clustering techniques that seem to be efficient in terms of frequency usage as long as the number of features used for *Model* generation remains low. On the other hand, [28] suggest the use of SVM for anomaly detection. SVM are well prepared for dealing with high dimensional data at the expense of a higher computational cost that might be not suitable for real-time detection.

As mentioned by [82], unsupervised anomaly detection approaches are suitable to deal with the intrusion detection problem as long as the number of attacks remains below the 2%. However, in practice this assumption is not always true. There are some situations in which for specific periods of time, the presence of intrusions could exceed the number of normal traffic records. For instance, when a new vulnerability is discovered and it has been widely announced, it is possible to find attacks exploiting this vulnerability encompassing a extremely high percentage of the network traffic.

Table 4. Comparative analysis of Anomaly-based Intrusion detection approaches

Authors	Model Generation Technique	Usage frequency	Model Acquisition/Adaptability	Features	Attack Type	Related Works
Porras and Valdes (1998)	**Statistical:** Chi-square-like	real*	**automatic**: readjust model with new attack-free patterns	**low and high-level:** packet, flow, payload	DoS, Probe, R2L, U2R	
Staniford et al. (2002)	**Statistical:** Naive Bayes Networks	real	**automatic**: readjust model with new attack-free patterns	**low-level**: packet, flow	Probe	
Lakhina et al. (2004)	**Statistical:** PCA	real*	**automatic**: readjust model with new attack-free patterns	**low-level**: flow	DoS, Probe	[17, 51, 50]
Mahoney and Chan (2003)	**Machine Learning:** rules learning and Markov models	batch	**automatic**: retraining with new attack-free patterns	**low and high-level:** packet, flow, payload	DoS, Probe, R2L, Worm	[71, 72, 70]
Lee and Stolfo (1998)	**Data mining**: Rules Learning and Frequent pattern count	batch	**automatic**: retraining with new attack-free patterns	**low and high-level:** packet, flow	non-specified	
Portnoy et al. (2002)	**Data mining**: unsupervised Clustering	batch	**automatic**: retraining with patterns containing a reduced amount of attacks	**low and high-level:** packet, flow,payload	Probe, R2L, DoS	[77, 93, 27]
Eskin et al. (2002)	**Data mining:** unsupervised SVM	batch	**automatic**: retraining with patterns containing a reduced amount of attacks	**low and high-level:** packet, flow payload	Probe, R2L, DoS	[60, 97]

0.5 (*) The use of flow information requires performing some sort of batch detection.

3.3.3. Feature Selection Analysis

If we consider the information in the fourth and fifth columns of Table 3 and Table 4, it is clear that there is a relation between the type of feature used by each approach and the kind of attacks detected. For instance, performing *R2L* detection without considering *payload* information such as a failed login attempt or anomalous HTTP requests (just to mention a few) it is extremely difficult. The previous relation explains why those approaches relying on *payload* information [86, 61, 82, 28, 80, 69] are capable of detecting *R2L* and in some cases even *U2R* attacks [86, 35, 80]. On the other hand, a common approach for detecting *Probes, Worms and DoS* attacks consists of using information regarding network *flows*. For example, recognizing a horizontal scan attack requires observing the number of connections from a host to different destinations in a predefined time window. In this regard, most of the discussed approaches capable of detecting *Probes* and *DoS* attacks except in [80, 92] use flows information for their *Traffic Features Generation* module.

Finally, another issue that should be considered is the relation between the selected features and the computational resources required for extracting such features. For example, obtaining high-level features from *flows* networks (such as number of connection from the same host) demands higher computational resources than just extracting low-level features available in the packet header. Moreover, since flow information such as bytes transferred or transfer rate are not directly available, obtaining such information requires processing packets inside a time window. In these sense, it can be said that those systems claiming real-time detection based on flow-based features [61, 81, 50], are actually performing some sort of batch detection. Clearly, a precise selection process must be conducted, considering the trade-off between the consumed computational resources and the attack detection accuracy provided by each type of feature.

4. Remaining Open Issues

The majority of the previously discussed works focus on the classification problem behind intrusion detection. If we considered the extremely precise results obtained by some approaches, we would say that the detection problem is near to be solved. Then, we should ask why none beyond pattern signature-based approach it is currently being used by network administrators. The fact is that previously analyzed works only cope with a subset of the problems that are essential to truly achieving intrusion detection, while not addressing the others. Issues like the still *high level of human interaction* and the *lack of model adjustment information* are critical to the detection process, specially if we consider that the ultimate goal of intrusion detection is to make security staff's life easier. In addition, a *proper traffic features identification* and the *lack of resource consumption information* are two other issues that should certainly be considered for an appropriate deployment on real networks. Finally, the *lack of public network traffic data* for proper evaluation of the different approaches is another issue that should be addressed by any further research made on this topic.

4.1. High Level of Human Interaction

All current approaches still need a high degree of human interaction during the model construction process. SNORT [86] requires expert knowledge for writing signature patterns. Similar is the case of P-BEST, for which the writing of a set of implication rules could demand a considerable human effort.

On the other hand, most of the current approaches aiming at automatically generating network traffic models still need a high level of human preprocessing of the input data. For instance, data mining and machine learning misuse-based approaches require network traffic data labeled as normal or intrusive, whereas anomaly-based approaches require traffic records labeled as normal, in other words attack-free traffic data. Even unsupervised approaches, yet to a lesser extent, require input data remains under some specific distribution, a situation that can only be guarantee by human experts. This need of human preprocessing is perhaps one of the major drawbacks in the deployment of those approaches aiming at automatically generating traffic models.

The intrusion detection research community have started to react to this issue providing the so-called hybrid approaches [10, 43]. Hybrid approaches usually combine well-established NIDS like SNORT with automatically generated traffic models techniques. Such combinations make the deployment of automatically generated models techniques more feasible. In addition, they seem to help reduce the required human preprocessing.

4.2. Lack of Model Adjustment Information

Most of the discussed approaches using automatic traffic models seem to be aware of the high network variability and provide methods for adapting themselves as needed. However, the appropriate time for performing such adjustments seems not to be analyzed enough. Approaches should be able to transfer their results into information that allow the network security staff to easily evaluate their possible course of actions for further system adjustment. In addition, more methods focused on determining when the traffic model is no longer representative of current network traffic might help the acceptance of these automatic approaches.

4.3. Proper Traffic Feature Identification

The majority of the current approaches address the detection problem from a broad point of view. These approaches rely on a large set of features to recognize several types of intrusion or, in some other cases, every possible type of intrusion. Unfortunately, using such broad strategies could complicate the already difficult problem of intrusion detection.

Alternatively, it seems more appropriate to focus on a per-attack detection strategy (i.e. specific to the kind of attack) and then analyze the adequate set of features capable of detecting it. Clearly, there is a relation between the kind of intrusion and the type of feature used for detecting such intrusion type. For instance, let's consider the Botnets detection problem. A major difference between Botnets and other detection problems consists of the use of a Command and Control communication channel (CC) for coordinating bots activities. We can infer that a proper identification of CC flows could eventually lead to detect infected machines on the network. In that case, the problem will be reduced to find just a concise set of features for proper CC flow characterization. In particular, this Botnet detection approach is currently being explored by members of the intrusion detection community [37].

This per-attack detection strategy however, brings about a new and important issue regarding the proper and efficient interaction between each one of these per-attack approaches. Possibly, the application of multi classifiers approaches and information fusion could provide some insights into this subject [90].

In addition, the size of the network and their associated security policies also have consequences when selecting network traffic features. As suggested by [31], aspects like the behavior of a feature over the time or the source of traffic features can vary according to network size. On the other hand, when considering network security policies, there could be certain activities considered normal in some environments but not in others. Recently, researchers have started to analyze the consequences of applying some of the reviewed techniques under different network sizes and network authorities [73].

In summary, for a better addressing of the problem, current approaches should select their features according to a set of attacks sharing similar behavior, considering local policies and explicitly defining the appropriate network size.

4.4. Lack of Resources Consumption Information

Determining the proper usage frequency for a given detection approach is crucial for analyzing its potential deployment on real networks. However, despite being a subject always present in surveys on intrusion detection, it is still difficult to establish the true usage frequency for many of the proposed approaches.

The problem is that only a few of the previously discussed intrusion detection approaches have analyzed the performance in terms of the computational resources required for generating the model as well as for evaluating a set of new network traffic records. As a result, it is difficult to establish the proper usage frequency of those approaches performing batch and real-time detection.

For instance, a batch detection approach could be able to perform detection every 5 minutes while another could be able to do it only every 12 hours. Since no measure is provided, it is difficult to establish the proper network conditions in which such approaches will be adequate. Even worse is the case of those approaches claiming ability to perform real-time detection. In many cases such claims have proved to be valid only under certain network conditions (bandwidth, throughput, etc.). Moreover, as stated in section 3.3.3 some of those approaches claiming real-time detection are actually performing batch detection.

This lack of resource consumption information could be one of the reasons why (with the exception of signature-based approaches) none of these approaches have been successfully deployed on real networks. Consequently, a better analysis about the needed computational resources could help in establishing the adequate usage frequency and therefore facilitating the deployment on real networks.

4.5. Lack of Public Network Traffic Data

Finally, another significant issue regarding intrusion detection is the lack of appropriate public data sets for evaluating the different approaches. Nowadays, the most commonly used data sets used for evaluation [62, 63] are almost 12 years old, which make them practically obsolete if we consider the fast evolution of the network security field. Current data sets should contain information on automatic attacks (e.g. Botnets), Peer-to-Peer traffic and distributed DoS attacks, among others. Moreover, since IPv6 based network have become a reality [103], security threats and attack types that can affect such kind of network should also be included.

There have been some efforts for providing a framework for dataset generation in a proper and duplicable way [99, 89]. However, in many cases, the research community continues evaluating its intrusion detection approaches using their own data without providing information about data set generation. A situation that seriously affects the principle of repeatability of experiments required for scientific research.

5. Conclusion

Several approaches have been proposed during the last 20 years of research in the intrusion detection field. All of these approaches aimed to facilitate the work of the network security staff providing some level of automation in the intrusion detection process. Certainly, such task can not be considered easy since the non-stationary behavior of network traffic along with the permanent growth of the network throughput.

Nowadays, NIDS most successful approaches are those based on pattern signatures describing known attacks behavior. However, since signatures describing such attacks have to be written by experts, the signature-based approach lacks of the required level of automation needed by the intrusion detection community.

In order to deal with the drawbacks regarding signature-based approaches, some novel techniques were applied to intrusion detection. Such techniques includes statistical methods, machine learning and data mining methods, among others. However, despite being successful in terms of accuracy detection, there are still several concerns to be considered for deploying these techniques on real life scenarios. Issues like the need of proper resources consumption information along with a more attack-specific feature selection have not been analyzed in depth. But above all these issues, perhaps the main reason for their failure when deployed on real network environment is that these automatic detection techniques still require a high degree of human interaction.

In summary, it can be said that most of the discussed techniques have shown being capable to obtain high accuracy levels in a more automatic way. However, to reach the objective of fully automatizing the intrusion detection process, issues like the labeling of network traffic and its associated resources should be more deeply analyzed. In addition, reducing the problem scale by considering more attack-specific oriented approaches could also aid to achieve the complete deployment of automatic detection techniques.

References

[1] Abadeh, M. S. and Habibi, J. (2010). A hybridization of evolutionary fuzzy systems and ant colony optimization for intrusion detection. *ISeCure, The ISC International Journal of Information Security*, 2(1):33–46.

[2] Abadeh, M. S., Habibi, J., Barzegar, Z., and Sergi, M. (2007a). A parallel genetic local search algorithm for intrusion detection in computer networks. *Engineering Applications of Artificial Intelligence*, 20:1058–1069.

[3] Abadeh, M. S., Habibi, J., and Lucas, C. (2007b). Intrusion detection using a fuzzy genetics-based learning algorithm. *Journal of Network and Computer Applications*, 30(1):414–428.

[4] Abadeh, M. S., Mohamadi, H., and Habibi, J. (2011). Design and analysis of genetic fuzzy systems for intrusion detection in computer networks. *Expert Systems with Applications*, 38(6):7067–7075.

[5] Abraham, A. and Grosan, C. (2006). Evolving intrusion detection systems. In Nedjah, N., Mourelle, L., and Abraham, A., editors, *Genetic Systems Programming*, volume 13 of *Studies in Computational Intelligence*, pages 57–79. Springer Berlin / Heidelberg.

[6] Ahmad, I., Abdullah, A. B., and Alghamdi, A. S. (2009). Artificial neural network approaches to intrusion detection: a review. In *Proceedings of the 8th Wseas international conference on Telecommunications and informatics*, pages 200–205, Stevens Point, Wisconsin, USA. World Scientific and Engineering Academy and Society (WSEAS).

[7] Anderson, D., Frivold, T., Tamaru, A., and Valdes, A. (1995). Next generation intrusion detection expert system (nides). Technical Report SRI-CSL-95-07, Computer Science Laboratory, SRI International, Menlo Park, California.

[8] Axelsson, S. (1998). Research in intrusion-detection systems: A survey. Technical Report 98–17, University of Technology, Goteborg, Sweden.

[9] Axelsson, S. (2000). The base-rate fallacy and the difficulty of intrusion detection. *ACM Transactions on Information and System Security*, 3:186–205.

[10] AydIn, M. A., Zaim, A. H., and Ceylan, K. G. (2009). A hybrid intrusion detection system design for computer network security. *Computers & Electrical Engineering*, 35(3):517 – 526.

[11] Banerjee, S., Grosan, C., and Abraham, A. (2005). IDEAS: Intrusion detection based on emotional ants for sensors. In *Proceedings. 5th International Conference on Intelligent Systems Design and Applications, 2005. ISDA '05.*, volume 0, pages 344–349, Los Alamitos, CA, USA. IEEE Computer Society.

[12] Bankovic, Z., Stepanovic, D., Bojanic, S., and Nieto-Taladriz, O. (2007). Improving network security using genetic algorithm approach. *Computers & Electrical Engineering*, 33(5-6):438–451.

[13] Bonabeau, E., Dorigo, M., and Theraulaz, G. (1999). *Swarm intelligence: from natural to artificial systems*. Oxford University Press, Inc., New York, NY, USA.

[14] Bonifacio, J., Cansian, A., De Carvalho, A., and Moreira, E. (1998). Neural networks applied in intrusion detection systems. In *Neural Networks Proceedings, 1998. IEEE World Congress on Computational Intelligence. The 1998 IEEE International Joint Conference on*, volume 1, pages 205–210.

[15] Boro, D., Nongpoh, B., and Bhattacharyya, D. K. (2012). Anomaly based intrusion detection using meta ensemble classifier. In *Proceedings of the Fifth International Conference on Security of Information and Networks*, SIN '12, pages 143–147, New York, NY, USA. ACM.

[16] Bridges, S. and Vaughn, R. (2000). Fuzzy data mining and genetic algorithms applied to intrusion detection. In *Proceedings of the 23rd National Information Systems Security Conference*, pages 13–31, Baltimore, MA., USA.

[17] Callegari, C., Gazzarrini, L., Giordano, S., Pagano, M., and Pepe, T. (2010). Detecting network anomalies in backbone networks. In Jha,

S., Sommer, R., and Kreibich, C., editors, *Recent Advances in Intrusion Detection*, volume 6307 of *Lecture Notes in Computer Science*, pages 490–491. Springer Berlin / Heidelberg.

[18] Cannady, J. (1998). Artificial neural networks for misuse detection. In *Proceedings of National Information Systems Security Conference*, pages 368–381, Arlington, VA, USA.

[19] Chandola, V., Banerjee, A., and Kumar, V. (2009). Anomaly detection: A survey. *ACM Computing. Surveys*, 41:15:1–15:58.

[20] Chen, C., Mabu, S., Yue, C., Shimada, K., and Hirasawa, K. (2009). Analysis of fuzzy class association rule mining based on genetic network programming. In *ICCAS-SICE*, 2009, pages 3480–3484.

[21] Chittur, A. (2001). Model generation for an intrusion detection system using genetic algorithms. *High School Honors Thesis, Ossining High School. In cooperation with Columbia Univ.*

[22] Cisco Systems (2008). Cisco IOS NetFlow configuration guide, release 12.4.

[23] Cohen, W. (1995). Fast effective rule induction. In *Proceedings of the 12th International Conference on Machine Learning*, pages 115–123. Morgan Kaufmann.

[24] Crosbie, M. and Spafford, G. (1995). Applying genetic programming to intrusion detection. In *Working Notes for the AAAI Symposium on Genetic Programming*, pages 1–8, Cambridge, MA.

[25] Debar, H., Dacier, M., and Wespi, A. (1999). Towards a taxonomy of intrusion-detection systems. *Computer Networks*, 31(8):805–822.

[26] Dickerson, J. and Dickerson, J. (2000). Fuzzy network profiling for intrusion detection. In *Fuzzy Information Processing Society, 2000. NAFIPS. 19th International Conference of the North American*, pages 301 –306.

[27] Ertöz, L., Eilertson, E., Lazarevic, A., Tan, P. N., Kumar, V., Srivastava, J., and Dokas, P. (2004). *MINDS - Minnesota Intrusion Detection System*, chapter 3. MIT Press.

[28] Eskin, E., Arnold, A., Prerau, M., Portnoy, L., and Stolfo, S. (2002). A geometric framework for unsupervised anomaly detection. In Barbara, D. and Jajodia, S., editors, *Advances in Information Security*, chapter 4. Springer.

[29] Florez, G., Bridges, S., and Vaughn, R. (2002). An improved algorithm for fuzzy data mining for intrusion detection. In *Fuzzy Information Processing Society, 2002. Proceedings. NAFIPS. 2002 Annual Meeting of the North American*, pages 457–462.

[30] Fowler, H. and Leland, W. (1991). Local area network characteristics, with implications for broadband network congestion management. *Selected Areas in Communications, IEEE Journal on*, 9(7):1139–1149.

[31] Garcia-Teodoro, P., Diaz-Verdejo, J., Macia-Fernandez, G., and Vazquez, E. (2009). Anomaly-based network intrusion detection: Techniques, systems and challenges. *Computers & Security*, 28(1-2):18 – 28.

[32] Giacinto, G., Perdisci, R., Rio, M. D., and Roli, F. (2008). Intrusion detection in computer networks by a modular ensemble of one-class classifiers. *Information Fusion*, 9(1):69–82. Special Issue on Applications of Ensemble Methods.

[33] Giacinto, G., Roli, F., and Didaci, L. (2003). Fusion of multiple classifiers for intrusion detection in computer networks. *Pattern Recognition Letters*, 24(12):1795–1803.

[34] Gomez, J. and Dasgupta, D. (2002). Evolving fuzzy classifiers for intrusion detection. In *Proceedings of the 2002 IEEE Workshop on Information Assurance*. New York: IEEE Computer Press.

[35] Gomez, J., Dasgupta, D., Nasraoui, O., and Gonzalez, F. (2002). Complete expression trees for evolving fuzzy classifier systems with genetic algorithms and application to network intrusion detection. In *Proceedings NAFIPS Fuzzy Information Processing Society 2002 Annual Meeting of the North American*, pages 469–474.

[36] Gong, R. H., Zulkernine, M., and Abolmaesumi, P. (2005). A software implementation of a genetic algorithm based approach to network intrusion detection. *Software Engineering, Artificial Intelligence, Networking*

and Parallel/Distributed Computing, International Conference on & Self-Assembling Wireless Networks, International Workshop on, 0:246–253. ISBN: 0-7695-2294-7.

[37] Gu, G., Perdisci, R., Zhang, J., and Lee, W. (2008). Botminer: clustering analysis of network traffic for protocol- and structure-independent botnet detection. In *Proceedings of the 17th conference on Security symposium, USENIX*, pages 139–154, Berkeley, CA, USA. USENIX Association. ISBN 978-1-931971-59-1.

[38] Gu, Y., McCallum, A., and Towsley, D. (2005). Detecting anomalies in network traffic using maximum entropy estimation. In *Proceedings of the 5th ACM SIGCOMM conference on Internet Measurement*, IMC '05, pages 32–32, Berkeley, CA, USA. USENIX Association. ISBN 1-59593-009-4.

[39] Guolong, C., Qingliang, C., and Wenzhong, G. (2007). A PSO-based approach to rule learning in network intrusion detection. In Cao, B.-Y., editor, *Fuzzy Information and Engineering*, volume 40 of *Advances in Intelligent and Soft Computing*, pages 666–673. Springer Berlin / Heidelberg. ISSN 1616-8658.

[40] Hansman, S. and Hunt, R. (2005). A taxonomy of network and computer attacks. *Computers & Security*, 24(1):31–43.

[41] He, J., Long, D., and Chen, C. (2007). An improved ant-based classifier for intrusion detection. *International Conference on Natural Computation*, 4:819–823. ISBN 0-7695-2875-9.

[42] Holland, J. H. (1992). *Adaptation in Natural and Artificial Systems: An Introductory Analysis with Applications to Biology, Control, and Artificial Intelligence*. The MIT Press.

[43] Hwang, K., Cai, M., Chen, Y., and Qin, M. (2007). Hybrid intrusion detection with weighted signature generation over anomalous internet episodes. *Dependable and Secure Computing, IEEE Transactions on*, 4(1):41–55.

[44] Igure, V. and Williams, R. (2008). Taxonomies of attacks and vulnerabilities in computer systems. *Communications Surveys Tutorials, IEEE*, 10(1):6–19.

[45] Javitz, H. and Valdes, A. (1991). The SRi IDES statistical anomaly detector. In *Research in Security and Privacy, 1991. Proceedings., 1991 IEEE Computer Society Symposium on*, pages 316–326.

[46] Jian-hua, Z. and Wei-hua, L. (2012). Intrusion detection based on bp neural network and genetic algorithm. In Liu, C., Wang, L., and Yang, A., editors, *Information Computing and Applications*, volume 308 of *Communications in Computer and Information Science*, pages 438–444. Springer Berlin Heidelberg.

[47] Kendall, K. (1999). A database of computer attacks for the evaluation of intrusion detection systems. Master's thesis. AAI3006082.

[48] Kittler, J., Hatef, M., Duin, R. P. W., and Matas, J. (1998). On combining classifiers. *IEEE Trans. Pattern Anal. Mach. Intell.*, 20:226–239.

[49] Kumar, P. A. R. and Selvakumar, S. (2011). Distributed denial of service attack detection using an ensemble of neural classifier. *Computer Communications*, 34(11):1328–1341.

[50] Lakhina, A., Crovella, M., and Diot, C. (2004a). Diagnosing network-wide traffic anomalies. *SIGCOMM Comput. Commun. Rev.*, 34:219–230.

[51] Lakhina, A., Crovella, M., and Diot, C. (2005). Mining anomalies using traffic feature distributions. *SIGCOMM Comput. Commun. Rev.*, 35:217–228.

[52] Lakhina, A., Papagiannaki, K., Crovella, M., Diot, C., Kolaczyk, E. D., and Taft, N. (2004b). Structural analysis of network traffic flows. In *Proceedings of the joint international conference on Measurement and modeling of computer systems*, SIGMETRICS '04/Performance '04, pages 61–72, New York, NY, USA. ACM.

[53] Lazarevic, A., Ertöz, L., Kumar, V., Ozgur, A., and Srivastava, J. (2003). A Comparative Study of Anomaly Detection Schemes in Network Intrusion Detection. In *Proceedings of the Third SIAM International Conference on Data Mining*.

[54] Lazarevic, A., Kumar, V., and Srivastava, J. (2005). Intrusion detection: A survey. In Kumar, V., Srivastava, J., and Lazarevic, A., editors,

Managing Cyber Threats, volume 5 of *Massive Computing*, pages 19–78. Springer US.

[55] Lee, W. and Stolfo, S. J. (1998). Data mining approaches for intrusion detection. In *Proceedings of the 7th conference on USENIX Security Symposium - Volume 7*, pages 6–6, Berkeley, CA, USA. USENIX Association. ISBN 978-1-931971-59-1.

[56] Lee, W., Wang, C., and Dagon, D. (2007). *Botnet Detection: Countering the Largest Security Threat (Advances in Information Security).* Springer-Verlag New York, Inc., Secaucus, NJ, USA.

[57] Lee, W. and Xiang, D. (2001). Information-theoretic measures for anomaly detection. In *Security and Privacy, 2001. S P 2001. Proceedings. 2001 IEEE Symposium on*, pages 130–143. ISSN 1081-6011.

[58] Leon, E., Nasraoui, O., and Gomez, J. (2004). Anomaly detection based on unsupervised niche clustering with application to network intrusion detection. In *Evolutionary Computation, 2004. CEC2004. Congress on*, volume 1, pages 502–508.

[59] Li, W. (2004). Using genetic algorithm for network intrusion detection. In *Proceedings of the United States Department of Energy Cyber Security Group 2004 Training Conference, Kansas City, Kansas*, pages 24–27. Department of Computer Science and Engineering. Mississippi State University, Mississippi State.

[60] Li, Y., Xia, J., Zhang, S., Yan, J., Ai, X., and Dai, K. (2012). An efficient intrusion detection system based on support vector machines and gradually feature removal method. *Expert Systems with Applications*, 39(1):424–430.

[61] Lindqvist, U. and Porras, P. (1999). Detecting computer and network misuse through the production-based expert system toolset (P-BEST). In *Security and Privacy, 1999. Proceedings of the 1999 IEEE Symposium on*, pages 146–161. ISSN 1081-6011.

[62] Lippmann, R., Fried, D., Graf, I., Haines, J., Kendall, K., McClung, D., Weber, D., Webster, S., Wyschogrod, D., Cunningham, R., and Zissman, M. (2000a). Evaluating intrusion detection systems: the 1998 DARPA

off-line intrusion detection evaluation. In *DARPA Information Survivability Conference and Exposition, 2000. DISCEX '00. Proceedings*, volume 2, pages 12 –26 vol.2. 640.

[63] Lippmann, R., Haines, J. W., Fried, D. J., Korba, J., and Das, K. (2000b). The 1999 DARPA off-line intrusion detection evaluation. *Computer Networks*, 34(4):579–595. Recent Advances in Intrusion Detection Systems.

[64] Liu, G., Yi, Z., and Yang, S. (2007). A hierarchical intrusion detection model based on the PCA neural networks. *Neurocomputing*, 70(7-9):1561 – 1568.

[65] Lu, W. and Traore, I. (2004). Detecting new forms of network intrusion using genetic programming. *Computational Intelligence*, 20(3):475–494.

[66] Luo, J. (1999). Integrating fuzzy logic with data mining methods for intrusion detection. *Master's thesis*, Department of Computer Science, Mississippi State University.

[67] Mahalanobis, P. C. (1936). On the generalised distance in statistics. In *Proceedings National Institute of Science*, India, volume 2, pages 49–55.

[68] Mahoney, M. and Chan, P. (2002a). Learning models of network traffic for detecting novel attacks. *Technical Report* CS-2002-08, Florida Institute of Technology.

[69] Mahoney, M. and Chan, P. (2003). Learning rules for anomaly detection of hostile network traffic. In *Data Mining, 2003. ICDM 2003. Third IEEE International Conference on*, pages 601–604.

[70] Mahoney, M. V. (2001). Detecting novel attacks by identifying anomalous network packet headers. *Technical Report* CS-2001-2, Florida Institute of Technology.

[71] Mahoney, M. V. and Chan, P. K. (2001). Phad: Packet header anomaly detection for identifying hostile network traffic. *Technical Report* CS-2001-4 2004, Florida Institute of Technology.

[72] Mahoney, M. V. and Chan, P. K. (2002b). Learning nonstationary models of normal network traffic for detecting novel attacks. In *Proceedings*

of the eighth ACM SIGKDD international conference on Knowledge discovery and data mining, KDD '02, pages 376–385, New York, NY, USA. ACM.

[73] Mehdi, S., Khalid, J., and Khayam, S. (2011). Revisiting traffic anomaly detection using software defined networking. In Sommer, R., Balzarotti, D., and Maier, G., editors, *Recent Advances in Intrusion Detection*, volume 6961 of *Lecture Notes in Computer Science*, pages 161–180. Springer Berlin / Heidelberg.

[74] Mukherjee, B., Heberlein, L., and Levitt, K. (1994). Network intrusion detection. *Network, IEEE*, 8(3):26–41.

[75] Mukkamala, S., Janoski, G., and Sung, A. (2002). Intrusion detection using neural networks and support vector machines. In *Neural Networks, 2002. IJCNN '02. Proceedings of the 2002 International Joint Conference on*, volume 2, pages 1702 –1707. ISSN 1098-7576.

[76] Patcha, A. and Park, J.-M. (2007). Network anomaly detection with incomplete audit data. *Computer Networks*, 51:3935–3955.

[77] Pathak, V. and Ananthanarayana, V. (2012). A novel multi-threaded k-means clustering approach for intrusion detection. In *Software Engineering and Service Science (ICSESS), 2012 IEEE 3rd International Conference on*, pages 757 –760. ISBN 978-1-4673-2007-8.

[78] Paxson, V. (1999). BRO: a system for detecting network intruders in real-time. *Computer Networks*, 31(23-24):2435–2463. 1455.

[79] Perdisci, R., Ariu, D., Fogla, P., Giacinto, G., and Lee, W. (2009). McPAD: A multiple classifier system for accurate payload-based anomaly detection. *Comput. Netw.*, 53(6):864–881.

[80] Porras, P. and Valdes, A. (1998). Live traffic analysis of TCP/IP gateways. In *Internet Society Symposium on Network and Distributed System Security, San Diego*.

[81] Porras, P. A. and Neumann, P. G. (1997). EMERALD: event monitoring enabling responses to anomalous live disturbances. In *Proceedings of the 20th National Information Systems Security Conference*, pages 353–365.

[82] Portnoy, L., Eskin, E., and Stolfo, S. (2001). Intrusion detection with unlabeled data using clustering. In *Proceedings of ACM CSS Workshop on Data Mining Applied to Security*, Philadelphia, PA.

[83] Ramadas, M., Ostermann, S., and Tjaden, B. (2003). Detecting anomalous network traffic with self-organizing maps. In Vigna, G., Kruegel, C., and Jonsson, E., editors, *Recent Advances in Intrusion Detection*, volume 2820 of *Lecture Notes in Computer Science*, pages 36–54. Springer Berlin / Heidelberg.

[84] Ramos, V. and Abraham, A. (2005). ANTIDS: Self organized ant-based clustering model for intrusion detection system. In Abraham, A., Dote, Y., Furuhashi, T., Köppen, M., Ohuchi, A., and Ohsawa, Y., editors, *Soft Computing as Transdisciplinary Science and Technology*, volume 29 of *Advances in Intelligent and Soft Computing*, pages 977–986. Springer Berlin / Heidelberg.

[85] Rehak, M., Pechoucek, M., Grill, M., Stiborek, J., Bartos, K., and Celeda, P. (2009). Adaptive multiagent system for network traffic monitoring. *Intelligent Systems, IEEE*, 24(3):16–25.

[86] Roesch, M. (1999). Snort - lightweight intrusion detection for networks. In *Proceedings of the 13th USENIX conference on System administration*, LISA '99, pages 229–238, Berkeley, CA, USA. USENIX Association.

[87] Sabater, J. and Sierra, C. (2005). Review on computational trust and reputation models. *Artificial Intelligence Review*, 24:33–60.

[88] Scarfone, K. and Mell, P. (2007). Guide to intrusion detection and prevetion systems (idps). *Technical Report* 800-94, National Institute of Standards and Technology. US. Departament of Commerce.

[89] Shiravi, A., Shiravi, H., Tavallaee, M., and Ghorbani, A. A. (2012). Toward developing a systematic approach to generate benchmark datasets for intrusion detection. *Computers & Security*, 31(3):357 – 374.

[90] Shoemaker, L. and Hall, L. O. (2011). Anomaly detection using ensembles. In *Proceedings of the 10th international conference on Multiple classifier systems*, MCS'11, pages 6–15, Berlin, Heidelberg. Springer-Verlag.

[91] Sindhu, S. S. S., Geetha, S., and Kannan, A. (2012). Decision tree based light weight intrusion detection using a wrapper approach. *Expert Systems with Applications*, 39(1):129–141.

[92] Staniford, S., Hoagland, J. A., and McAlerney, J. M. (2002). Practical automated detection of stealthy portscans. *Journal of Computer Security*, 10:105–136.

[93] Su, M. (2011). Using clustering to improve the KNN-based classifiers for online anomaly network traffic identification. *Journal of Network and Computer Applications*, 34(2):722–730.

[94] Sung, A. and Mukkamala, S. (2003). Identifying important features for intrusion detection using support vector machines and neural networks. In *Applications and the Internet, 2003. Proceedings. 2003 Symposium on*, pages 209 – 216. ISBN 0-7695-1872-9.

[95] SURICATA (2011). Suricata intrusion detection system.

[96] Vollmer, T., Alves-Foss, J., and Manic, M. (2011). Autonomous rule creation for intrusion detection. In *Computational Intelligence in Cyber Security (CICS), 2011 IEEE Symposium on*, pages 1–8. ISBN 978-1-4244-9905-2.

[97] Wang, F., Qian, Y., Dai, Y., and Wang, Z. (2010). A model based on hybrid support vector machine and self-organizing map for anomaly detection. *Communications and Mobile Computing, International Conference on*, 1:97–101. ISBN 978-0-7695-3989-8.

[98] Wong, M. L. and Leung, K. S. (2000). *Data Mining Using Grammar-Based Genetic Programming and Applications*. Genetic Programming Series. Kluwer Academic Publishers, Norwell, MA, USA.

[99] Wright, C., Connelly, C., Braje, T., Rabek, J., Rossey, L., and Cunningham, R. (2010). Generating client workloads and high-fidelity network traffic for controllable, repeatable experiments in computer security. In Jha, S., Sommer, R., and Kreibich, C., editors, *Recent Advances in Intrusion Detection*, volume 6307 of *Lecture Notes in Computer Science*, pages 218–237. Springer Berlin / Heidelberg.

[100] Xu, Q., Pei, W., Yang, L., and Zhao, Q. (2006). An intrusion detection approach based on understandable neural network trees. *International Journal of Computer Science and Network Security*, 6(11):229–234.

[101] Ye, N., Li, X., and Emran, S. (2000). Decision tree for signature recognition and state classification. In *Proceedings of IEEE Systems, Man and Cybernetics Information Assurance & Security Workshop*.

[102] Yi, Y., Wu, J., and Xu, W. (2011). Incremental svm based on reserved set for network intrusion detection. *Expert Systems with Applications*, 38(6):7698–7707.

[103] Zagar, D., Grgic, K., and Rimac-Drlje, S. (2007). Security aspects in ipv6 networks - implementation and testing. *Computers & Electrical Engineering*, 33(5-6):425 – 437. Security of Computers &amp Networks.

[104] Zhao, J.-h. and Li, W.-h. (2012). Improvement intrusion detection based on svm. In Liu, C., Wang, L., and Yang, A., editors, *Information Computing and Applications*, volume 308 of *Communications in Computer and Information Science*, pages 53–60. Springer Berlin Heidelberg.

[105] Zhou, C. V., Leckie, C., and Karunasekera, S. (2010). A survey of coordinated attacks and collaborative intrusion detection. *Computers & Security*, 29(1):124 – 140.

Index

A

B

C

D

E

F

M

N

O

P

Q

R

S

T

U

V

W

Y

Z